LONDON MATHEMATICAL SOCIETY LECTURE NOTE SERIES

Managing Editor: Professor Endre Süli, Mathematical Institute,
University of Oxford, Woodstock Road, Oxford OX2 6GG, United Kingdom

The titles below are available from booksellers, or from Cambridge University Press at
www.cambridge.org/mathematics

London Mathematical Society Lecture Note Series: 462

Zeta and *L*-Functions of Varieties and Motives

BRUNO KAHN

CNRS, Institut de Mathématiques de Jussieu-Paris Rive Gauche

CAMBRIDGE
UNIVERSITY PRESS

CAMBRIDGE
UNIVERSITY PRESS

University Printing House, Cambridge CB2 8BS, United Kingdom

One Liberty Plaza, 20th Floor, New York, NY 10006, USA

477 Williamstown Road, Port Melbourne, VIC 3207, Australia

314–321, 3rd Floor, Plot 3, Splendor Forum, Jasola District Centre, New Delhi – 110025, India

79 Anson Road, #06–04/06, Singapore 079906

Cambridge University Press is part of the University of Cambridge.

It furthers the University's mission by disseminating knowledge in the pursuit of education, learning, and research at the highest international levels of excellence.

www.cambridge.org
Information on this title: www.cambridge.org/9781108703390
DOI: 10.1017/9781108691536

Originally published in French as *Fonctions zêta et L de variétés et de motifs* by Calvage et Mounet in 2018

© Bruno Kahn 2018

First published in English by Cambridge University Press in 2020

English language translation © Cambridge University Press 2020

A catalogue record for this publication is available from the British Library.

ISBN 978-1-108-70339-0 Paperback

Contents

Contents

Introduction

Gauss described number theory as the queen of mathematics. Indeed, the amount of mathematics invented for arithmetic reasons is truly astonishing. To name just a few examples:

- A large part of complex analysis (Cauchy, Riemann, Weierstraß, Hadamard, Hardy–Littlewood, ...);
- The theory of divisors (Kummer, Dedekind) and of ideals (E. Noether);
- The theory of Riemann surfaces (Riemann, ...);
- The non-analytic version of the Riemann–Roch theorem for curves (F. K. Schmidt);
- The refoundation of Italian algebraic geometry on the basis of commutative algebra (Weil);[1]
- The theory of abelian varieties (Weil);
- Part of the theory of linear representations of finite groups (Artin, Brauer);
- A large part of homological algebra: group cohomology, the theory of sheaves on a general site (Cartan, Eilenberg, Serre, Tate, Grothendieck, ...);
- Part of the theory of schemes (Grothendieck);
- The development of étale cohomology (M. Artin, Grothendieck, Verdier, Deligne...) and then crystalline cohomology (Grothendieck, Berthelot, Ogus, Bloch, Deligne, Illusie, ...);
- Part of the theory of derived categories (Grothendieck, Verdier);
- The six operations formalism (Grothendieck);
- Monodromy theory in the world of schemes (Grothendieck, Serre, Deligne, Katz, ...);

and, of course, the theory of motives!

---

[1] van der Waerden and Zariski also participated in this movement, for different reasons.

Zeta and L-functions are the meeting point of these theories: they crown the queen of mathematics. It is remarkable that a function whose definition is as simple as

$$\zeta(s) = \sum_{n=1}^{\infty} \frac{1}{n^s}$$

has played such a deep role, admitting vast generalisations which have shaped the evolution of number theory up to the present day; yet it remains the object of a conjecture whose analogue for varieties over finite fields was famously proved by Weil and Deligne, but whose original case due to Riemann remains an unapproachable mystery.

The text that follows is derived from a master's level course given at Jussieu in 2013. I have tried to give an exposition of part of the known results on zeta and L-functions, but also of the complex issues surrounding them, in an *ontogenetic* way: ontogenesis describes the progressive development of an organism from its conception to its mature form. With this goal in mind, I have scattered the text with quotations and commentaries which, I hope, will offer the reader a small window on the history of ideas in this field.

After having recalled in Chapter 1 the classical results (and hypothesis) on the Riemann zeta function, I introduce zeta functions of $\mathbf{Z}$-schemes of finite type in the second chapter, which is essentially dedicated to the proof of the Riemann hypothesis for curves over a finite field. To my great regret, I was not able to do justice to Weil's proof of the Castelnuovo–Severi inequality in [Wei2]: to re-transcribe it in the language of schemes would have led me too far astray (see [64]). I only include in § 2.9.5 an idea of the proof, and treat the easy case of curves of genus 1 (due to Hasse). For the general case, I followed everyone else in giving the proofs of Mattuck–Tate and Grothendieck, which rely on an *a priori* weaker inequality; as a consolation prize, I compare the two inequalities in § 2.9.7 and show that we can recover the first one using the second and the additivity of Pic^τ (the numerically trivial divisors). A second proof of the Castelnuovo–Severi inequality via abelian varieties will be found in the next chapter (§ 3.2.2). It explains the first proof and puts it in a new light.

Chapter 3 is dedicated to the Weil conjectures. They are all proven, except for the hardest one: the "Riemann hypothesis". I also give an overview of Dwork's p-adic proof of the rationality of zeta functions of varieties over a finite field (obtained before the development of Grothendieck's cohomological methods!).

Chapter 4 returns to more elementary mathematics, introducing Dirichlet, Hecke, and Artin L-functions. I give a proof of Dirichlet's theorem on arithmetic progressions, by the method expounded by Serre in [Ser3]; it would however be a shame, as Pierre Charollois remarked to me, to omit Dirichlet's original method, which gave additional information and anticipated the analytic class number formulae (see Remark 4.22(2)). I then introduce the two generalisations of Dirichlet's L-functions: those of Hecke and Artin. I state without proof Hecke's main theorem: existence of an analytic continuation and a functional equation (Theorem 4.60), and then explain how Artin and Brauer derived the same results for non-abelian L-functions (Theorem 4.69).

The next chapter is the *pièce de résistance* of the text. It starts by introducing the approximate idea of Hasse–Weil L-functions, and ends by describing their precise definition due to Serre [113]. In the mean time, I explain the fundamental contributions of Grothendieck and Deligne: rationality and the functional equation for L-functions of l-adic sheaves in characteristic p, with essentially complete proofs; the theory of weights; and some theorems of Deligne on the Riemann hypothesis (the last of Weil's conjectures), this time without proof. Notably, one will find in § 5.4.4 an exposition of functional equations of the L-functions developed by Grothendieck in [CorrGS, letter from 30-9-64], and one will find in Theorem 5.58 a more precise statement and fairly complete parts of the proof of Grothendieck and Deligne's theorem on the rationality and the functional equation of Hasse–Weil L-functions in characteristic > 0, confirming a conjecture of Serre (Conjecture 5.57) in this case.

The last chapter is dedicated to motives and their zeta functions. I limited myself to an elementary case: that of pure motives of Grothendieck, associated to smooth projective varieties over a finite field. One can go much further using the triangulated categories of motives introduced by Voevodsky and developed by Ivorra, Ayoub, and Cisinski–Déglise, but this would go beyond the scope of the book (see [63]). I explain how this viewpoint considerably clarifies how Weil cohomologies are used to prove rationality and the functional equation, and I do not resist the pleasure of applying this theory to prove a somewhat forgotten theorem of Weil: Artin's conjecture for non-abelian L-functions in positive characteristic (§ 6.15).

Finally, two appendices give supplements from categorical algebra. I also scattered the text with exercises, without attempting anything systematic.

Aside from Hecke theory, the automorphic side of the story is left essentially unaddressed; I contented myself with brief allusions here and there.

There still remains one question: what is the difference between a zeta function and an L-function? Morally, an L-function is a zeta function "with coefficients" (in a sheaf, in a representation, ...), and we recover a zeta function by taking trivial coefficients (cf. § 5.3). The situation becomes more complicated when we consider the Euler products associated by Hasse and Weil to a curve over a number field... Another problem is which terminology one should use in the case of motives. For pure motives, I have chosen that of zeta functions, which seems well established. It would probably be simpler to unify the terminology by dropping one of the two notations, but this is obviously delicate, for reasons of tradition.

I have benefited a great deal from previous expositions of the theory, from which I have borrowed much: these expositions would be too numerous to list here comprehensively, so I simply refer the reader to the bibliography. I thank the audience from the course, and in particular Matthieu Rambaud, for pointing out a number of misprints. Finally, I thank Joseph Ayoub, Pierre Charollois, Luc Illusie, Amnon Neeman, Ram Murty, and Jean-Pierre Serre for pertinent comments on this text.

Guide to the reader

This book is not a systematic exposition of a theory of zeta and L-functions – which does not exist, contrary to class field theory for example. I have chosen primarily to highlight the history of ideas. Thus, while certain proofs are complete, in others I prefer to emphasise the main ideas rather than give a complete but excessively technical proof; when such a proof is already available, I refer to expositions containing the missing parts.

Finally, there are cases where I have provided details that cannot be found in the literature. Here are some of them:

- § 5.2: a discussion of the notion of good reduction in relation to Hasse–Weil zeta functions (see in particular Proposition 5.4).
- § 5.4: Grothendieck's functional equation, described in [CorrGS, letter from 30-9-64] (Theorems 5.29 and 5.30).
- Theorem 5.58, deducing from this functional equation the one described by Serre in [113, 4.1, example a)], with a precision that follows from "Weil II" (Deligne [30]).
- § 6.8: the theory of specialisation for pure motives.
- §§ 6.12 and 6.13: the zeta function of an endomorphism of a motive, its rationality, and its functional equation (see also § A.2.8 for an abstract categorical version, reproducing that of [60, I] in a slightly weaker form).
- § 6.15: a motivic proof of Artin's conjecture in positive characteristic.
- § A.2.4: the notion of a "category with suspension".

I have also recalled (without proof) several standard results, such as the Riemann–Roch theorem, the theory of abelian varieties, and the basics of algebraic number theory, for which there are already excellent expositions.

Finally, I assume some prerequisites, varying from one chapter to another:

- Chapters 1, 2 and §§ 4.1, 4.2: the basics of complex analysis.
- § 1.10 and Chapter 4: the basics of algebraic number theory, they can be found for example in the book of Lang [Lan2], in Cassels–Fröhlich [CF], among others.
- Chapters 2 and 3: the basics of algebraic geometry, as can be found for example in the book of Hartshorne [Har2]; the basics of intersection theory, as can be found in the book of Fulton [Ful].
- § 4.3: the basics of general topology.
- §§ 4.4 and 4.6: the basics of the theory of linear representations of finite groups, as in Serre's book [Ser1].
- Chapter 5: much more serious foundations of the theory of schemes and étale cohomology, as developed in [SGA4, SGA4$\frac{1}{2}$, SGA5] and [Mil]. This chapter is undoubtedly the hardest to read.
- Chapter 6: the basics of categorical algebra, as in the book of Mac Lane [Mcl].

Preface to the English edition

This version is a direct translation of the French version; it differs only by a few corrections and the addition of four exercises in Chapter 6. The numbering has also been changed.

1

The Riemann zeta function

The Riemann zeta function is defined to be

$$\zeta(s) = \sum_{n=1}^{\infty} \frac{1}{n^s}.$$

This function was first studied by Euler, and later by Riemann.

1.1 A bit of history

1.1.1 Euler

Euler established:

- $\zeta(2) = \frac{\pi^2}{6}$ (the Basel problem, 1735).
- The *Euler product*: $\zeta(s) = \prod_{p \text{ prime}} (1 - p^{-s})^{-1}$ (thesis, 1737).

He also obtained the more general formula

$$\zeta(2n) = (-1)^{n+1} \frac{B_{2n}(2\pi)^{2n}}{2(2n)!},$$

where the B_n denote the *Bernoulli–Seki numbers*[1].

1.1.2 Riemann ([95], 1859)

Riemann studied the function $\zeta(s)$ for $s \in \mathbf{C}$, establishing that:

---

[1] Seki Takakazu or Seki Kôwa was a Japanese mathematician and contemporary of Bernoulli, who discovered "Bernoulli numbers" in the same period as the latter, i.e. towards the end of the seventeenth century.

- It converges absolutely for $\Re(s) > 1$.
- It admits a meromorphic extension to $\mathbf{C}$, with a simple pole at $s = 1$ with residue $= 1$.
- It satisfies a functional equation (essentially conjectured by Euler in 1749): if we let $\Lambda(s) = \pi^{-s/2}\Gamma(s/2)\zeta(s)$, then

$$\Lambda(s) = \Lambda(1 - s).$$

- He also conjectured the Riemann hypothesis on the distribution of the zeros of $\zeta(s)$.

The functional equation gives a formula for the value of $\zeta(s)$ at negative integers (guessed by Euler):

$$\zeta(-n) = -\frac{B_{n+1}}{n+1} \quad (n > 0), \quad \zeta(0) = -\frac{1}{2}.$$

Riemann's motivation was the *prime number theorem* (on the distribution of prime numbers). This last result had been anticipated by Čebyšev (1848–1850) and was finally proved by Hadamard and de la Vallée Poussin (1896).

We will now prove some of these results. Part of the following exposition is borrowed from Serre [Ser3].

1.2 Absolute convergence

Proposition 1.1 *The function $\zeta(s)$ converges absolutely and uniformly on every compact set in the domain $\Re(s) > 1$, in which it defines a holomorphic function. It diverges at $s = 1$.*

Proof For both convergence and divergence, if $s = \sigma + it$ with $\sigma, t \in \mathbf{R}$, then we have $|n^{-s}| = n^{-\sigma}$, so we may suppose that $s \in \mathbf{R}$. We then use the following fact.

Lemma 1.2 (integral test) *Let $f : \mathbf{R}^+ \to \mathbf{R}^+$ be a decreasing function. Then $\sum_{n=1}^{\infty} f(n) < \infty \iff \int_1^{\infty} f(t)dt < \infty$.*

Proof We have the inequalities

$$f(n) = \int_n^{n+1} f(n)dt \geq \int_n^{n+1} f(t)dt \geq \int_n^{n+1} f(n+1)dt = f(n+1),$$

that is to say,

$$\int_{n-1}^{n} f(t)dt \geq f(n) \geq \int_n^{n+1} f(t)dt,$$

and we take the sum. $\qquad\qquad\qquad\qquad\qquad\qquad\qquad\qquad\qquad\qquad\qquad\square$

This proof also gives the inequality

$$\left|\sum_{n=N}^{\infty} n^{-s}\right| \leq \int_{N-1}^{\infty} t^{-\Re(s)}dt = \frac{(N-1)^{1-\Re(s)}}{\Re(s)-1},$$

whence uniform convergence; holomorphy follows from the next lemma, applied to the relevant partial sums. $\qquad\square$

Lemma 1.3 *Let U be an open subset of $\mathbf{C}$, and let $f_n : U \to \mathbf{C}$ be a sequence of holomorphic functions converging uniformly to f on every compact set in U. Then f is holomorphic on U, and the derivatives f_n' of f_n converge uniformly to the derivative f' of f on every compact set.*

Proof Let D be a closed disc in U with boundary ∂D, oriented counterclockwise, and let $z_0 \in \overset{\circ}{D}$. We apply the Cauchy formula to f_n:

$$f_n(z_0) = \frac{1}{2i\pi} \int_{\partial D} \frac{f_n(z)}{z-z_0}dz.$$

By passing to the limit, we obtain

$$f(z_0) = \frac{1}{2i\pi} \int_{\partial D} \frac{f(z)}{z-z_0}dz,$$

which shows that f is holomorphic on U, establishing the first part of the lemma. The formula

$$f'(z_0) = \frac{1}{2i\pi} \int_{\partial D} \frac{f(z)}{(z-z_0)^2}dz$$

gives the second part in the same manner. $\qquad\square$

1.3 The Euler product

Proposition 1.4 (Euler) *We have a factorisation*

$$\zeta(s) = \prod_p \frac{1}{1-p^{-s}},$$

where the product, taken over all prime numbers, converges absolutely for $\Re(s) > 1$.

For the proof, say that a function $a : \mathbf{N} - \{0\} \to \mathbf{C}$ is *multiplicative* (resp. *completely multiplicative*) if it satisfies the identity $a(mn) = a(m)a(n)$ for all coprime m, n (resp. for all m, n). The proposition follows from the next lemma applied to the function $a(n) = n^{-s}$.

Lemma 1.5 *Let a be a multiplicative function. Then the following conditions are equivalent:*

(i) $\displaystyle\sum_{n=1}^{\infty} |a(n)| < +\infty$;

(ii) $\displaystyle\prod_{p} \left(1 + |a(p)| + \cdots + |a(p^m)| + \cdots \right) < +\infty.$

If these conditions are satisfied, then

$$\sum_{n=1}^{\infty} a(n) = \prod_{p} \left(1 + a(p) + \cdots + a(p^m) + \cdots \right),$$

and if a is completely multiplicative:

$$\sum_{n=1}^{\infty} a(n) = \prod_{p} \frac{1}{1 - a(p)}.$$

Proof Suppose that (i) holds. In particular, $\sum a(p^m)$ converges absolutely for every p. For $x \in \mathbf{N}$, we then have

$$\sum_{n \in E(x)} a(n) = \prod_{p<x} \sum_{m} a(p^m),$$

where

$$E(x) = \{n \in \mathbf{N} - \{0\} \mid \text{all prime factors of } n \text{ are } < x\}.$$

Hence,

$$\left| \sum_{n=1}^{\infty} a(n) - \prod_{p<x} \sum_{m} a(p^m) \right| = \left| \sum_{n \notin E(x)} a(n) \right| \leq \sum_{n \geq x} |a(n)|$$

which shows that the infinite product converges to $\sum_{n=1}^{\infty} a(n)$. By applying this to $|a(n)|$, we see that the convergence is absolute, thus (ii) is satisfied.

Now suppose that (ii) holds. Then

$$\sum_{n<x} |a(n)| \leq \sum_{n \in E(x)} |a(n)| = \prod_{p<x} \left(1 + |a(p)| + \cdots + |a(p^m)| + \ldots \right),$$

thus (i) is satisfied. □

Corollary 1.6 *The function $\zeta(s)$ does not vanish for $\Re(s) > 1$.*   □

Exercise 1.7 (Euler) Show that $\sum_{p \in \mathcal{P}} 1/p$ diverges.

1.4 Formal Dirichlet series

The notion of formal Dirichlet series is very useful for doing formal computations before thinking about questions of convergence.

Definition 1.8 (cf. [Ell, ch. 3, A4]) We define a *formal Dirichlet series* (with complex coefficients) to be an expression of the form

$$f = \sum_{n \geq 1} \frac{a_n}{n^s},$$

where n runs over the positive integers, and the a_n are complex numbers. If $f = \sum_{n \geq 1} \frac{a_n}{n^s}$ and $g = \sum_{n \geq 1} \frac{b_n}{n^s}$ are two formal Dirichlet series, we define their sum and their product as

$$f + g = \sum_{n \geq 1} \frac{a_n + b_n}{n^s},$$

$$fg = \sum_{n \geq 1} \frac{c_n}{n^s},$$

where $c_n = \sum_{pq=n} a_p b_q$.

Formal Dirichlet series form a unital commutative ring: the algebra of formal Dirichlet series with complex coefficients, denoted Dir($\mathbf{C}$).

Remark 1.9 The algebra Dir($\mathbf{C}$) is nothing other than the *total algebra*, sometimes also called the *large algebra* (following the terminology used by Bourbaki), of the multiplicative monoid $\mathbf{N} - \{0\}$ [BA3, ch. III, § 2, n° 10]. By writing each positive integer as a product of its prime factors, we see that Dir($\mathbf{C}$) is isomorphic to the algebra of formal power series with complex coefficients on a countably infinite set of variables (cf. [BLie2, ex. 2 of the appendix]).

Definition 1.10 Let $f = \sum_{n \geq 1} \frac{a_n}{n^s}$ be a formal Dirichlet series. If $f \neq 0$, we define the *order* of f, denoted by $\omega(f)$, to be the smallest integer n such that $a_n \neq 0$. If $f = 0$, we set $\omega(f) = +\infty$.

Proposition 1.11 *The subsets* $\{f \mid \omega(f) \geq N\}$ *of* Dir(**C**) *are ideals of* Dir(**C**). *These subsets define a topology on* Dir(**C**) *under which* Dir(**C**) *is a complete topological ring.* □

Corollary 1.12

(1) *The sequence* $(f_n)_{n\in\mathbf{N}}$ *is summable in* Dir(**C**) *if and only if* $\lim_{n\to\infty} \omega(f_n) = +\infty$.
(2) *The sequence* $(1 + f_n)_{n\in\mathbf{N}}$, *where* $\omega(f_n) > 1$ *for all* n, *is multiplicable in* Dir(**C**) *if and only if* $\lim_{n\to\infty} \omega(f_n) = +\infty$. □

Lemma 1.5 gives:

Proposition 1.13 *Let* $a : \mathbf{N}-\{0\} \to \mathbf{C}$ *be a completely multiplicative function, and let* $\alpha \in \mathbf{R}$. *In the formal identity*

$$\sum_{n\geq 1} \frac{a(n)}{n^s} = \prod_p \left(1 - \frac{a(p)}{p^s}\right)^{-1},$$

the left-hand side converges absolutely for $\Re(s) > \alpha$ *if and only if the right-hand side converges absolutely for* $\Re(s) > \alpha$.

Exercise 1.14

(a) Prove the following identity in Dir(**C**):

$$\zeta(s)^{-1} = \sum_{n=1}^{\infty} \frac{\mu(n)}{n^s}, \quad \text{with}$$

$$\mu(n) = \begin{cases} (-1)^k & \text{if } n \text{ is the product of } k \text{ distinct prime numbers} \\ 0 & \text{otherwise.} \end{cases}$$

(The function $\mu(n)$ is called the *Möbius function*.)
(b) Deduce from (a) the *Möbius inversion formula*: if $b_n = \sum_{d|n} a_d$, then $a_n = \sum_{d|n} \mu(d) b_{n/d}$.

More generally, let $f : [1, +\infty[\to \mathbf{C}$ be a function and $F(x) = \sum_{n\leq x} f\left(\frac{x}{n}\right)$. Then we have $f(x) = \sum_{n\leq x} \mu(n) F\left(\frac{x}{n}\right)$.

1.5 Extension to $\Re(s) > 0$; the pole and residue at $s = 1$

Proposition 1.15 *The function $\zeta(s)$ extends meromorphically to the domain $\Re(s) > 0$, with a simple pole at $s = 1$ with residue 1.*

To see this we need the following lemmas.

Lemma 1.16 *Let $(\alpha, \beta) \in \mathbf{R}^2$ be such that $0 < \alpha < \beta$. Let $z \in \mathbf{C}$ with $\Re(z) > 0$. Then we have*

$$\left| e^{-\alpha z} - e^{-\beta z} \right| \le \left| \frac{z}{x} \right| (e^{-\alpha x} - e^{-\beta x}),$$

where $x = \Re(z)$.

Proof We write $e^{-\alpha z} - e^{-\beta z} = z \int_\alpha^\beta e^{-tz} dt$, and by passing to absolute values:

$$\left| e^{-\alpha z} - e^{-\beta z} \right| \le |z| \int_\alpha^\beta e^{-tx} dt = \left| \frac{z}{x} \right| (e^{-\alpha x} - e^{-\beta x}). \qquad \square$$

Lemma 1.17 (Abel's lemma) *Let (a_n) and (b_n) be sequences of complex numbers. Define $A_{m,p} = \sum_{n=m}^{p} a_n$ and $S_{m,m'} = \sum_{n=m}^{m'} a_n b_n$. Then we have*

$$S_{m,m'} = \sum_{n=m}^{m'-1} A_{m,n}(b_n - b_{n+1}) + A_{m,m'} b_{m'}. \qquad \square$$

Proof of Proposition 1.15. Consider

$$\zeta_2(s) = \sum_{n=1}^{\infty} \frac{(-1)^{n+1}}{n^s}.$$

We apply Lemma 1.17 with $a_n = (-1)^{n+1}$, $b_n = n^{-s}$; Lemma 1.16 gives the inequality $\left| b_n - b_{n+1} \right| \le \left| \frac{s}{\sigma} \right| (n^{-\sigma} - (n+1)^{-\sigma})$, with $\sigma = \Re(s) > 0$, therefore

$$\left| \sum_{n=m}^{m'-1} \frac{(-1)^{n+1}}{n^s} \right| \le \left| \frac{s}{\sigma} \right| (m^{-\sigma} - m'^{-\sigma}) + m'^{-\sigma} \le (\left| \frac{s}{\sigma} \right| + 1) m^{-\sigma}.$$

We conclude from this that ζ_2 converges for $\Re(s) > 0$. In addition,

$$\zeta(s) - \zeta_2(s) = 2^{1-s} \zeta(s) \implies \zeta_2(s) = (1 - 2^{1-s}) \zeta(s).$$

It follows that the function $(s-1)\zeta(s)$ extends meromorphically to $\Re(s) > 0$. At $s = 1$, we have $\zeta_2(s) = \log 2$. L'Hôpital's rule then gives

$$\lim_{s \to 1} (s-1)\zeta(s) = 1.$$

More generally, for $r \in \mathbf{N} - \{0, 1\}$ consider

$$\zeta_r(s) = \frac{1}{1^s} + \frac{1}{2^s} + \cdots + \frac{1}{(r-1)^s} - \frac{r-1}{r^s} + \frac{1}{(r+1)^s} + \cdots$$

We see right away that the partial sums of the coefficients of ζ_r are bounded by $r - 1$. Reasoning as above, ζ_r is analytic for $\Re(s) > 0$. Furthermore, we have

$$\left(1 - \frac{1}{r^{s-1}}\right)\zeta(s) = \zeta_r(s). \tag{1.5.1}$$

Equation (1.5.1) for $r = 2$ shows that a pole $s \neq 1$ of $\zeta(s)$ must satisfy $2^{s-1} = 1$, i.e. $s = \frac{2i\pi k}{\log 2} + 1$ for an integer $k \neq 0$. Similarly, by using Equation (1.5.1) for $r = 3$, we see that $s = \frac{2i\pi \ell}{\log 3} + 1$ for ℓ an integer $\neq 0$. But this implies $3^k = 2^\ell$, which is impossible. $\qquad\square$

1.6 The functional equation

Recall the gamma function (defined by Euler in 1730):

$$\Gamma(s) = \int_0^\infty t^s e^{-t} \frac{dt}{t}.$$

This integral converges for $\Re(s) > 0$; we have $\Gamma(1) = 1$ and the functional equation

$$\Gamma(s+1) = s\Gamma(s). \tag{1.6.1}$$

It follows that $\Gamma(n) = (n-1)!$ for $n \in \mathbf{N} - \{0\}$, and that $\Gamma(s)$ extends meromorphically to the whole complex plane with a simple pole at $s = -n$ for each $n \in \mathbf{N}$, with residue $(-1)^n/n!$. In fact, we also have *Euler's reflection formula*:

$$\Gamma(s)\Gamma(1-s) = \frac{\pi}{\sin(\pi s)} \tag{1.6.2}$$

and the *multiplication theorem*:

$$\prod_{k=0}^{m-1} \Gamma\left(s + \frac{k}{m}\right) = (2\pi)^{\frac{m-1}{2}} m^{\frac{1}{2}-ms} \Gamma(ms). \tag{1.6.3}$$

Now set

$$\xi(s) = \frac{s(s-1)}{2}\pi^{-s/2}\Gamma(s/2)\zeta(s).$$

Theorem 1.18 (Riemann)

(1) *The function $\xi(s)$ extends holomorphically to the whole complex plane, with functional equation*

$$\xi(1-s) = \xi(s).$$

(2) *The function $\zeta(s)$ extends meromorphically to the whole complex plane, with functional equation*

$$\zeta(1-s) = \frac{2}{(2\pi)^s} \cos\left(\frac{\pi s}{2}\right) \Gamma(s)\zeta(s).$$

It has a simple zero at $s = -n$ for each even integer $n > 0$ (we call these zeros "trivial"), and has no other zeros outside of the "critical strip" $0 \le \Re(s) \le 1$.

(The function $\xi(s)$ differs from the function $\Lambda(s)$ of § 1.1.2 by a factor of $\frac{s(s-1)}{2}$: their functional equations are equivalent, and ξ has the advantage of being entire.)

Proof The formulae (1.6.2) and (1.6.3) reduce (2) to (1). I will follow the proof given in Ellison [Ell, proof of th. 5.2]:

For each s satisfying $\Re(s) > 0$ and for every integer $n > 0$, we have

$$\Gamma\left(\frac{s}{2}\right)(\pi n^2)^{-\frac{s}{2}} = \int_0^\infty t^{\frac{s}{2}-1} e^{-\pi n^2 t}\, dt.$$

Suppose $\Re(s) > 1$, then by summing over $n > 0$:

$$\Gamma\left(\frac{s}{2}\right)\pi^{-\frac{s}{2}}\zeta(s) = \sum_{n=1}^\infty \int_0^\infty t^{\frac{s}{2}-1} e^{-\pi n^2 t}\, dt.$$

Since the series on the right converges for s real and > 1, we can interchange the integral and the sum, which gives:

$$\Gamma\left(\frac{s}{2}\right)\pi^{-\frac{s}{2}}\zeta(s) = \int_0^\infty \sum_{n=1}^\infty t^{\frac{s}{2}-1} e^{-\pi n^2 t}\, dt = \int_0^\infty t^{\frac{s}{2}-1}\omega(t)\, dt$$

with $\omega(t) = \sum_{n=1}^\infty e^{-\pi n^2 t}$. By writing $\int_0^\infty = \int_0^1 + \int_1^\infty$ in the term on the right, it suffices to prove the formula

$$1 + 2\omega(t) = \frac{1}{\sqrt{t}}\left(1 + 2\omega\left(\frac{1}{t}\right)\right) \tag{1.6.4}$$

which would imply

$$\xi(s) = \frac{1}{2} + \frac{s(s-1)}{2}\int_1^\infty \left(t^{\frac{s}{2}} + t^{\frac{1-s}{2}}\right)\omega(t)\frac{dt}{t}.$$

(Since $\omega(t) = O(e^{-\pi t})$, the term on the right of this identity extends to the whole complex plane and is invariant under $s \mapsto 1 - s$.)

The functional equation (1.6.4) is the case $z = 0$ of the identity

$$\sum_{n=-\infty}^{+\infty} e^{-\pi(n+z)^2 t} = \frac{1}{\sqrt{t}} \sum_{n=-\infty}^{+\infty} e^{-\pi n^2/t + 2i\pi nz},$$

where the right-hand side is the Fourier expansion of the left-hand side, which is periodic with period 1 in z (cf. [Ell, proof of th. 5.1]). □

Remark 1.19 There is a much more elementary proof of the analytic continuation of $\zeta(s)$, in the spirit of Proposition 1.15, by exploiting the identity

$$(1 - 2^{1-s})\zeta(s) = \sum_{n=1}^{\infty} \frac{\Gamma(s+n)}{\Gamma(s)n!} \frac{\zeta(n+s)}{2^{n+s}}$$

(cf. [Ell, ch. 2, ex. 2.3]). Unfortunately, this proof does not appear to extend to Dedekind zeta functions. . .

1.7 The Riemann hypothesis

The results of the preceding section allow us to formulate:

Conjecture 1.20 (The Riemann hypothesis) *The nontrivial zeros of the function* $\zeta(s)$ *lie on the "critical line"* $\Re(s) = \frac{1}{2}$.

Why is the Riemann hypothesis important? Firstly because of the "explicit formula" linking the zeros of $\zeta(s)$ to the distribution of the prime numbers.

For $x \in \mathbf{R}^+$, set

$$\pi(x) = |\{p \mid p \text{ prime}, p \leq x\}|$$

and

$$\Pi(x) = \sum_{n=1}^{\infty} \frac{1}{n} \pi(x^{1/n}),$$

so that

$$\pi(x) = \sum_{n=1}^{\infty} \mu(n) \frac{1}{n} \Pi(x^{1/n}),$$

where μ is the Möbius function. We also set

$$\mathrm{Li}(z) = \int_0^z \frac{dt}{\log t} \quad \text{(logarithmic integral).}^2$$

Theorem 1.21 (Riemann, 1859) *We have*

$$\mathrm{Li}(x) - \sum_\rho \mathrm{Li}(x^\rho) - \log(2) + \int_x^\infty \frac{dt}{t(t^2-1)\log t}$$

$$= \lim_{\varepsilon \to 0} \frac{\Pi(x-\varepsilon) + \Pi(x+\varepsilon)}{2},$$

where the (non-absolutely convergent) sum on the left is over the nontrivial zeros of $\zeta(s)$, ordered by increasing absolute value of the imaginary parts.

Theorem 1.21 implies following result.

Theorem 1.22 (von Koch [136], Schoenfeld [103]) *The Riemann hypothesis holds if and only if*

$$|\pi(x) - \mathrm{Li}(x)| < \frac{1}{8\pi} x^{1/2} \log x \ \text{ for } x \geq 2657. \tag{1.7.1}$$

Here is another explicit formula, somewhat more tractable [Ell, ch. 5, th. 5.9]: let

$$Z(s) = -\frac{\zeta'(s)}{\zeta(s)} = \sum_{n=1}^\infty \frac{\lambda(n)}{n^s},$$

where

$$\lambda(n) = \begin{cases} \log p & \text{if } n \text{ is of the form } p^\nu, \\ 0 & \text{otherwise} \end{cases}$$

(known as the von Mangoldt function)[3]. Set

$$\psi(x) = \sum_{n \leq x} \lambda(n) = \sum_{p^\nu \leq x} \log p$$

and

$$\psi_0(x) = \lim_{\varepsilon \to 0} \frac{1}{2} \left(\psi(x+\varepsilon) + \psi(x-\varepsilon) \right) = \begin{cases} \psi(x) & \text{if } x \neq p^\nu, \\ \psi(x) - \frac{1}{2} \log p & \text{if } x = p^\nu. \end{cases}$$

---

[2] Some care is required to give a meaning to this divergent integral: it defines a multivalued function with branch points at 0 and 1.

[3] It is traditionally written $\Lambda(n)$, but this notation conflicts with the one we have adopted for the completed zeta function. See p. 7.

Theorem 1.23 *For all x > 1, we have*

$$\psi_0(x) = x + Z(0) - \frac{1}{2}\log(1 - x^{-2}) - \lim_{T \to \infty} \sum_{|\Im(\rho)| < T} \frac{x^{\rho}}{\rho},$$

where ρ runs over the nontrivial zeros of $\zeta(s)$.

For a considerable generalisation to the Hecke L-functions, see Weil [142].

Several arithmetic statements have been proved by reasoning along the following lines: the statement is true if the Riemann hypothesis is true, and on the other hand, it is true if the Riemann hypothesis is false. The most famous such statement is the finiteness of the set of imaginary quadratic fields of a given class number (Heilbronn [51]); Siegel finally gave a proof [118] in 1935 which did not involve the Riemann hypothesis.

1.8 Results and approaches

All the nontrivial zeros of $\zeta(s)$ that have been found lie on the critical line. Here are further results in that direction.

Theorem 1.24

(1) **(Hardy [47], 1914)** *The function $\zeta(s)$ has infinitely many zeros on the critical line.*
(2) **(Selberg, Levinson, Conrey [20], 1989)** *At least 2/5 of the nontrivial zeros lie on the critical line.*

Theorem 1.25

(1) **(Hadamard [45], de la Vallée Poussin [23], 1896)** *The function $\zeta(s)$ does not vanish on the line $\Re(s) = 1$.*
(2) **(Ford [41], 2002)** *If $s = \sigma +$ it is a nontrivial zero of $\zeta(s)$ with $|t| \geq 3$, then*

$$\sigma \leq 1 - \frac{1}{57.54(\log|t|)^{2/3}(\log\log|t|)^{1/3}}.$$

There have been numerous attempts to prove the Riemann hypothesis, which have so far all been unsuccessful. Several attempts have elaborated on an idea of Hilbert and Pólya: find a self-adjoint operator u on a Hilbert space such that the nontrivial zeros of $\zeta(s)$ correspond to the eigenvalues of u.

1.9 The prime number theorem

The prime number theorem is a weak version of (1.7.1).

Theorem 1.26 (Hadamard [45], de la Vallée Poussin [23], 1896) *We have*

$$\pi(x) \sim \frac{x}{\log x}, \quad x \to +\infty.$$

(Note that the function $\mathrm{Li}(x)$ is asymptotic to $x/\log x$ as x tends to infinity.)

The proof relies essentially on Theorem 1.25 (1): see [Ell, ch. 2] for a proof of this type, and [Ell, ch. 3] for an "elementary" proof. Conversely, we can deduce Theorem 1.25 (1) from the prime number theorem [Ell, ch. 3, § 4.4]: this situation is analogous to that of Theorem 1.22.

1.10 Dedekind zeta functions

If K is a number field, we define

$$\zeta_K(s) = \sum_{\mathfrak{A}} \frac{1}{N(\mathfrak{A})^s},$$

where $\mathfrak{A}$ runs through the ideals of the ring of integers O_K and $N(\mathfrak{A}) := [O_K : \mathfrak{A}]$ is the norm of $\mathfrak{A}$: this is the *Dedekind zeta function* of K. It has the following properties:

Euler product

$$\zeta_K(s) = \prod_{\mathfrak{P}} \frac{1}{1 - N(\mathfrak{P})^{-s}},$$

where $\mathfrak{P}$ runs through the prime ideals of O_K. This converges absolutely for $\Re(s) > 1$.

Absolute convergence for $\Re(s) > 1$.

Simple pole at $s = 1$.

Functional equation (Hecke, 1920, see th. 4.60). Set

$$\Gamma_{\mathbf{R}}(s) = \pi^{-s/2}\Gamma(s/2), \quad \Gamma_{\mathbf{C}}(s) = 2(2\pi)^{-s}\Gamma(s) \qquad (1.10.1)$$

$$\Lambda_K(s) = |d_K|^{s/2} \, \Gamma_{\mathbf{R}}(s)^{r_1} \Gamma_{\mathbf{C}}(s)^{r_2} \zeta_K(s),$$

where d_K is the absolute discriminant of K and (r_1, r_2) is the signature of K (number of real and complex embeddings). Then

$$\Lambda_K(s) = \Lambda_K(1 - s).$$

Residue at $s = 1$: by the functional equation, it suffices to give the leading term of $\zeta_K(s)$ at $s = 0$. We have

$$\text{ord}_{s=0}\, \zeta_K(s) = r_1 + r_2 - 1 =: r, \quad \lim_{s \to 0} s^{-r} \zeta_K(s) = -\frac{h_K R_K}{w_K},$$

where $h_K = |Cl(O_K)|$ is the class number of K, w_K is the number of roots of unity of K, and R_K is the *regulator*.

The proof of the factorisation as an Euler product is the same as that for the Riemann zeta function. Let us show that the convergence is absolute for $\Re(s) > 1$: by Proposition 1.13 we may suppose that s is real and work with the Euler product. Let p be a prime number: if $\mathfrak{P} \mid p$, then

$$\frac{1}{1 - N(\mathfrak{P})^{-s}} \leq \frac{1}{1 - p^{-s}}$$

since $N(\mathfrak{P})$ is a power of p. If $n = [K : \mathbf{Q}]$, then there are at most n prime ideals $\mathfrak{P}$ dividing p, hence

$$\prod_{\mathfrak{P}\mid p} \frac{1}{1 - N(\mathfrak{P})^{-s}} \leq \frac{1}{(1 - p^{-s})^n},$$

and

$$\zeta_K(s) \leq \zeta(s)^n,$$

which concludes the proof.

The other statements are much more difficult (Dedekind, Hecke, Tate: [Lan2], [126]).

Exercise 1.27 Using Formula (1.6.3), check the identity

$$\Gamma_{\mathbf{C}}(s) = \Gamma_{\mathbf{R}}(s)\Gamma_{\mathbf{R}}(s + 1).$$

2

The zeta function of a Z-scheme of finite type

2.1 A bit of history

References: Roquette [98], Serre [116], Osserman [89], Audin [12].

2.1.1 Emil Artin's thesis (1921)

Following a suggestion from his thesis advisor Herglotz, Artin develops the theory of zeta functions of quadratic function fields (over $\mathbf{F}_p(t)$) in analogy with the $\zeta_K(s)$ for K a quadratic number field. He proves the zeta functions to be rational (in p^{-s}) and formulates the "Riemann hypothesis" in this context, which he then checks numerically for several examples.

2.1.2 The theses of Sengenhorst, F.K. Schmidt and Rauter

These authors transfer most of the theorems known at the time for number fields to function fields of one variable over a finite field.

2.1.3 Friedrich Karl Schmidt and zeta functions [101]

Schmidt proves the Riemann–Roch theorem for function fields K of one variable over an arbitrary perfect field k (following Dedekind–Weber on $\mathbf{C}$). When k is finite, he generalises Artin's theory of zeta functions from a "birational" point of view (without choosing a ring of integers in K), and uses Riemann–Roch to prove both the functional equation of $\zeta_K(s)$ and its rationality. To this end, he proves that K always admits a divisor of degree 1.

2.1.4 Helmut Hasse and Max Deuring

Hasse gives *two* proofs of the Riemann hypothesis for function fields of an elliptic curve E. The first unpublished proof, for $p > 3$, uses a uniformisation method (lifting to characteristic zero), as well as the theory of complex multiplication.[1] The second proof [49] uses properties of the endomorphism ring of E, in particular the Frobenius endomorphism. Deuring simplifies Hasse's proof using algebraic correspondences [34].

2.1.5 André Weil

Weil proves the Riemann hypothesis for arbitrary curves C over a finite field, introducing geometric language in the process. The notes [138] and [139] are incomplete reports; they are completed in [Wei2] and [Wei3], after Weil re-founded algebraic geometry for this exact purpose [Wei1]. He gives two proofs: the first, in [Wei2], relies on the theory of correspondences on C and is reproduced below; the second, in [Wei3, n° 48], essentially uses the Tate module of the Jacobian of C, and applies more generally to an abelian variety: I offer a version of this proof in § 3.2.

Weil formulates the *Weil conjectures* for arbitrary varieties over a finite field [140].

2.2 Elementary properties of $\zeta(X, s)$

Reference: Serre [112].

Definition 2.1 Let X be a scheme of finite type over $\mathbf{Z}$. We set

$$\zeta(X, s) = \prod_{x \in X_{(0)}} \frac{1}{1 - N(x)^{-s}},$$

where $X_{(0)}$ denotes the set of closed points of X and $N(x)$ is the cardinality of the residue field $\kappa(x)$ (which is finite by hypothesis on X).

(This definition only depends on the reduced structure of X, or even on the "atomisation" $X_{(0)}$ of X.)

---

[1] See [98, II, 5.3] for a description of this proof: Roquette explains that it generalises (independently) an earlier proof due to Herglotz for the case of the function field of the lemniscate.

Example 2.2 For $X = \operatorname{Spec} O_K$, where K is a number field, we recover the Dedekind zeta function $\zeta_K(s)$ of § 1.10.

Proposition 2.3

(1) *The product defining $\zeta(X, s)$ is multiplicable in* $\operatorname{Dir}(\mathbf{C})$.
(2) *If $X_{(0)} = \coprod_{r=1}^{\infty}(X_r)_{(0)}$, where X_r are subschemes, we have*

$$\zeta(X, s) = \prod_{r=1}^{\infty} \zeta(X_r, s).$$

In particular,

$$\zeta(X, s) = \prod_{p} \zeta(X_p, s)$$

where X_p is the fibre of $X \to \operatorname{Spec} \mathbf{Z}$ at the prime number p.
(3) *(F.K. Schmidt for a curve) If X is a $\mathbf{F}_q$-scheme, then we have*

$$\zeta(X, s) = \exp\left(\sum_{n=1}^{\infty} |X(\mathbf{F}_{q^n})|\, \frac{q^{-ns}}{n}\right).$$

(4) *Let $\mathbf{A}_X^1$ (resp. $\mathbf{P}_X^1$) denote the affine (resp. projective) line on X. Then we have*

$$\zeta(\mathbf{A}_{\mathbf{F}_q}^1, s) = \frac{1}{1 - q^{1-s}}, \quad \zeta(\mathbf{P}_{\mathbf{F}_q}^1, s) = \frac{1}{(1 - q^{-s})(1 - q^{1-s})}.$$

(5) *We have the identity*

$$\zeta(\mathbf{A}_X^1, s) = \zeta(X, s - 1).$$

Proof (1) As X is of finite type over $\mathbf{Z}$, there are only a finite number of $x \in X_{(0)}$ such that $N(x) \leq N$, where N is a given integer. It follows that the product is multiplicable (cf. Corollary 1.12 (2)). In the process, we obtain the formula

$$\zeta(X, s) = \sum_{c \in Z_0(X)^+} \frac{1}{N(c)^s} \qquad (2.2.1)$$

as a formal Dirichlet series, where $Z_0(X)$ is the group of 0-cycles of X, $Z_0(X)^+$ is the submonoid of positive cycles [2], and

$$N(c) = \prod N(x)^{n_x}$$

if $c = \sum n_x x$.

---

[2] I prefer using the adjective "positive" to "effective"; the latter can then be reserved for motives without risk of confusion. This corresponds in fact with Weil's terminology [Wei2].

(2) is clear. For (3), denote the degree of a closed point x by $\deg(x) = [\kappa(x) : \mathbf{F}_q]$. We have

$$\log \zeta(X, s) = \sum_{x \in X_{(0)}} -\log(1 - N(x)^{-s}) = \sum_{x \in X_{(0)}} \sum_{m=1}^{\infty} \frac{N(x)^{-ms}}{m}$$

$$= \sum_{m=1}^{\infty} \sum_{x \in X_{(0)}} \frac{N(x)^{-ms}}{m} = \sum_{m=1}^{\infty} \sum_{x \in X_{(0)}} \frac{q^{-m \deg(x)s}}{m}$$

$$= \sum_{n=1}^{\infty} \sum_{\deg(x)|n} \deg(x) \frac{q^{-ns}}{n},$$

and it remains to observe that $\left| X(\mathbf{F}_{q^n}) \right| = \sum_{\deg(x)|n} \deg(x)$.

For (4), we apply (3) which gives

$$\zeta(\mathbf{A}^1_{\mathbf{F}_q}, s) = \exp \left(\sum_{n=1}^{\infty} \left| \mathbf{A}^1(\mathbf{F}_{q^n}) \right| \frac{q^{-ns}}{n} \right) = \exp \left(\sum_{n=1}^{\infty} q^n \frac{q^{-ns}}{n} \right) = \frac{1}{1 - q^{1-s}},$$

and

$$\zeta(\mathbf{P}^1_{\mathbf{F}_q}, s) = \zeta(\mathbf{F}_q, s)\zeta(\mathbf{A}^1_{\mathbf{F}_q}, s) = \frac{1}{(1 - q^{-s})(1 - q^{1-s})}.$$

Finally, for (5) we apply (2), which gives

$$\zeta(\mathbf{A}^1_X, s) = \prod_{x \in X_{(0)}} \zeta(\mathbf{A}^1_x, s),$$

from which we derive the formula using (4). □

Remark 2.4 Point (3) of Proposition 2.3 shows why zeta functions are interesting: they encode the number of rational points over finite fields. It also shows their limitations: two varieties with the same number of points under reduction mod $\mathfrak{p}$ for all $\mathfrak{p}$ have the same zeta function. Thus, $\zeta(\mathbf{P}^1_{\mathbf{Z}[1/n]}, s) = \zeta(C_{\mathbf{Z}[1/n]}, s)$ if C is an anisotropic conic over $\mathbf{Q}$ with the equation $x_0^2 = ax_1^2 + bx_2^2$, where a, b are integers prime to n and $C_{\mathbf{Z}[1/n]}$ is the smooth projective $\mathbf{Z}[1/n]$-scheme with the same equation. (Indeed, every conic over a finite field has a rational point due to the Chevalley–Warning theorem.) Another example, in equal characteristic, is that of two isogenous abelian varieties over a finite field. See also Proposition 5.6. For examples (of rings of integers) of non-isomorphic number fields having the same zeta function, see, for example, [92, 33].

Theorem 2.5 *The function* $\zeta(X, s)$ *converges absolutely (as an infinite sum or an infinite product) for* $\Re(s) > \dim X$.

Proof We proceed as follows (cf. [112, proof of th. 1]).

(1) If $X = X_1 \cup X_2$, where X_1 and X_2 are closed and X_2 is irreducible, Proposition 2.3 (1) gives the formula

$$\zeta(X, s) = \frac{\zeta(X_1, s)\zeta(X_2, s)}{\zeta(X_1 \cap X_2, s)},$$

so, by induction on $\dim X$, the statement for X_1 and X_2 implies the statement for X. We can therefore assume X irreducible.

(2) If U is an open subset of X with complement Z, we have similarly that

$$\zeta(X, s) = \zeta(U, s)\zeta(Z, s)$$

so, by induction on $\dim X$ again, the statement is equivalent for X and U. This reduces the problem to the case of affine (and integral) X.

(3) If $f : X \to Y$ is a finite flat morphism and the statement holds for Y, then it holds for X. Indeed, if d is the degree of f, the same reasoning as in § 1.10 gives $\zeta(X, s) \leq \zeta(Y, s)^d$ for $s \in (n, +\infty)$ where $n = \dim Y$.

(4) By the Noether normalisation lemma, we arrive at the case $X = \mathbf{A}_{\mathbf{Z}}^n$ or $X = \operatorname{Spec} \mathbf{A}_{\mathbf{F}_q}^n$ depending on whether X is flat or not over $\mathbf{Z}$[3]: in the second case, Proposition 2.3 (5) gives

$$\zeta(X, s) = \frac{1}{1 - q^{n-s}},$$

which converges absolutely for $\Re(s) > n = \dim X$; in the first case, the same proposition gives

$$\zeta(X, s) = \zeta(s - n),$$

which converges absolutely for $s > n + 1 = \dim X$, by Proposition 1.1.

$\square$

2.3 The case of a curve over a finite field: the statement

If X is a $\mathbf{F}_q$-scheme of finite type, by Proposition 2.3 (3) we can write:

$$\zeta(X, s) = Z(X, q^{-s})$$

---

[3] More precisely, in the first case an affine open of X is finite over $\mathbf{A}_U^n$, where U is an open in $\operatorname{Spec} \mathbf{Z}$; this brings us to the case of $\mathbf{A}_{\mathbf{Z}}^n$, as in (2).

with $Z(X, t) \in \mathbf{Q}[[t]]$. The same result gives

$$Z(X, t) = \exp\left(\sum_{n=1}^{\infty} \nu_n \frac{t^n}{n}\right), \quad \nu_n = |X(\mathbf{F}_{q^n})|,$$

and the proof of Proposition 2.3 (1) gives another expression

$$Z(X, t) = \sum_{c \in Z_0(X)^+} t^{\deg(c)} = \sum_{n \geq 0} b_n t^n \tag{2.3.1}$$

where $b_n = |\{c \in Z_0(X)^+ : \deg(c) = n\}|$.

Warning 2.6 The function $\zeta(X, s)$ is "absolute", while $Z(X, t)$ depends on the choice of the base field $\mathbf{F}_q$: if X is an $\mathbf{F}_{q'}$-scheme of finite type with $q' = q^r$, then $Z(X/\mathbf{F}_q, t) = Z(X/\mathbf{F}_{q'}, t^r)$.

Theorem 2.7 *Let C be a geometrically connected, smooth projective curve, of genus g over $\mathbf{F}_q$. Then*

$$Z(C, t) = \frac{P(t)}{(1-t)(1-qt)},$$

where $P \in \mathbf{Z}[t]$ is a polynomial of degree $2g$ with constant term 1 whose inverse roots have absolute value $\sqrt{q}$ for each embedding of $\bar{\mathbf{Q}}$ in $\mathbf{C}$. We also have the functional equation

$$Z(C, 1/qt) = q^{1-g}t^{2-2g}Z(C, t).$$

(The inverse roots of P are the numbers $\alpha_1, \ldots, \alpha_{2g}$ such that

$$P(t) = \prod_{i=1}^{2g}(1 - \alpha_i t).)$$

Corollary 2.8 (Riemann hypothesis) *The zeros of $\zeta(C, s)$ all lie on the critical line $\Re(s) = 1/2$.*

2.4 Strategy of the proof of Theorem 2.7

(i) Rationality and the functional equation follow from the Riemann–Roch theorem.

(ii) The Riemann hypothesis is proved by bounding the $a_n := 1 + q^n - \nu_n$. More precisely:

$$|a_n| \leq 2g\sqrt{q^n}. \tag{2.4.1}$$

In what follows, we will use "curve over k" to mean "geometrically connected, smooth projective curve over (the field) k".

2.5 Review of divisors

Let X be a smooth k-variety. A *divisor* on X is a finite linear combination $D = \sum_i n_i Z_i$, where Z_i are (irreducible) subvarieties of X of codimension 1 and n_i are integers; if the n_i are ≥ 0, we say that D is *positive*. We denote the group of divisors on X by $\mathrm{Div}(X)$, and the submonoid of positive divisors by $\mathrm{Div}(X)^+$. If X is a curve, $\mathrm{Div}(X)$ coincides with the group of 0-cycles $Z_0(X)$.

A divisor is *principal* if it is of the form $(f) = \sum v_Z(f)Z$, where f is a nonzero rational function on X, the Z are irreducible subvarieties of X of codimension 1 and $v_Z(f)$ is the valuation of f at the local ring of Z (which is a discrete valuation ring). The quotient of $\mathrm{Div}(X)$ by the subgroup of principal divisors is denoted $\mathrm{Pic}(X)$: this is the *Picard group* of X.

If X is a smooth projective curve, we let $\deg(D) = \sum n_i$: this is the *degree* of D. (More generally, the degree is defined for every 0-cycle on a smooth projective variety.) We have $\deg((f)) = 0$ for all rational functions $f \in k(X)^*$, hence we have a map $\deg : \mathrm{Pic}(X) \to \mathbf{Z}$; we set $\mathrm{Pic}^0(X)$ to be its kernel.

2.6 The Riemann–Roch theorem

Recall the statement of the Riemann–Roch theorem (cf. [Har2, ch. IV, § 1]) for a curve C of genus g over a field k.

Let F be the function field of C. If D is a divisor on C, we write $\mathcal{L}(D)$ for the corresponding invertible sheaf; then $H^0(C, \mathcal{L}(D))$ is a k-vector space of finite dimension $l(D)$ [Har2, ch. II, th. 5.19]. By definition, $\mathcal{L}(D)$ is a subsheaf of the constant sheaf F and we have an isomorphism [Har2, ch. II, prop. 7.7]

$$H^0(C, \mathcal{L}(D)) \simeq L(D) := \{f \in F^* \mid D + (f) \in \mathrm{Div}(C)^+\} \cup \{0\}.$$

Therefore $l(D)$ depends only on the class of D in $\mathrm{Pic}(C)$.

We shall write Ω^1 for the sheaf of Kähler differential forms: it is locally free of rank 1. We denote by K a divisor corresponding to Ω^1 (a canonical divisor). The Riemann–Roch theorem is the formula

$$l(D) - l(K - D) = \deg(D) + 1 - g. \tag{2.6.1}$$

In particular, we can deduce the following:

$$l(K) = g, \quad \deg(K) = 2g - 2, \tag{2.6.2}$$
$$l(D) = \deg(D) + 1 - g \text{ if } \deg(D) > 2g - 2 \tag{2.6.3}$$

(cf. [Har2, ch. IV, ex. 1.3.3 and 1.3.4]).

(In Hartshorne, k is assumed to be algebraically closed, but we reduce immediately to this case since the $l(D)$ are invariant under base extension.)

Exercise 2.9

(a) Show that if C is genus zero, then it is isomorphic to a smooth conic in $\mathbf{P}^2$. (Apply (2.6.3) to $D = -K$ by using the fact that D is very ample, cf. [Har2, ch. IV, cor. 3.2 (b)].)

(b) If in addition C has a rational point, it is isomorphic to $\mathbf{P}^1$.

(c) Show that any non-degenerate quadratic form in 3 variables on $\mathbf{F}_q$ admits a nontrivial zero if q is odd. (Reduce to an affine equation of type $x^2 = ay^2 + b$, and count the number of values of x^2 and of $ay^2 + b$.)

(d) Show that all quadratic forms φ in 3 variables over a perfect field of characteristic 2 admit a nontrivial zero. (Start by showing that φ is of the form $ax^2 + xy + by^2 + cz^2$.)

(e) Prove (2.4.1) when C is of genus zero.

2.7 Rationality and the functional equation (F.K. Schmidt)

We return to the case that interests us: $k = \mathbf{F}_q$. What follows is a re-transcription of [Wei2, nos 18–19].

Lemma 2.10 *Let $D \in Z_0(C)$. Then the number of positive divisors rationally equivalent to D is equal to*

$$\frac{q^{l(D)} - 1}{q - 1}.$$

Proof Indeed, by definition this number is the cardinality of the linear system $|D| = \mathbf{P}(L(D))$ associated to D. □

Let $\delta > 0$ be the greatest common divisor of the degrees of the nonzero elements of $Z_0(C)^4$; choose a divisor D_0 of degree δ. Let ν be an integer such that $\nu\delta \geq g$, and choose a maximal set $D_1, \ldots, D_h$ of positive divisors of degree $\nu\delta$, pairwise not linearly equivalent.

---

[4] We will see below that $\delta = 1$, but it is not necessary to know this now.

Lemma 2.11 *Every divisor D of degree $v\delta$ is rationally equivalent to precisely one D_i, $1 \le i \le h$.*

Proof Following (2.6.1), one has $l(D) > 0$, which implies that D is rationally equivalent to a positive divisor. □

Let D be a positive nonzero divisor: we have $\deg(D) = n\delta$, $n > 0$. Then $D - (n - v)D_0$ is of degree $v\delta$, so we may apply Lemma 2.11. Taking into account Lemma 2.10, this gives

$$b_{n\delta} = \sum_{i=1}^{h} \frac{q^{l(D_i + (n-v)D_0)} - 1}{q - 1}.$$

But, for $n\delta > 2g - 2$, (2.6.1) gives $l(D_i + (n - v)D_0) = n\delta + 1 - g$, so that

$$b_{n\delta} = h \frac{q^{n\delta + 1 - g} - 1}{q - 1}.$$

Therefore (2.3.1) tells us that $(q - 1)Z(C, t)$ is the sum of a polynomial and the series

$$h \sum_{n\delta > 2g-2} (q^{n\delta + 1 - g} - 1)t^{n\delta} = h \left(q^{1-g} \frac{(qt)^{(\rho+1)\delta}}{1 - (qt)^{\delta}} - \frac{t^{(\rho+1)\delta}}{1 - t^{\delta}} \right)$$

where $\rho = \inf\{n \mid n\delta \ge 2g - 2\}$. Thus, $Z(C, t) \in \mathbf{Q}(t)$.

We now remark that $\delta \mid 2g - 2$ (using (2.6.2)), so

$$\rho = \frac{2g - 2}{\delta}.$$

We can now state more precisely that $(q - 1)Z(C, t) = F(t) + hR(t)$, with

$$F(t) = \sum_{n=0}^{\rho} \sum_{i=1}^{h} q^{l(D_i + (n-v)D_0)} t^{n\delta}, \quad R(t) = \frac{1}{t^{\delta} - 1} + q^{1-g} \frac{(qt)^{(\rho+1)\delta}}{1 - (qt)^{\delta}}.$$

$$(2.7.1)$$

A direct calculation gives

$$R(1/qt) = q^{1-g} t^{2-2g} R(t).$$

To address $F(t)$, we re-use the Riemann–Roch formula: together with Lemma 2.11, it implies that the families $(l(K - D_i - (n - v)D_0))_{1 \le i \le h}$ and $(l(D_i + (\rho - n - v)D_0))_{1 \le i \le h}$ are the same up to permutation. We deduce that $F(t)$, and therefore $Z(C, t)$, satisfies the same functional equation as $R(t)$.

2.8 The Riemann hypothesis: reduction to (2.4.1) (Hasse, Schmidt, Weil)

Suppose (2.4.1) is known; set $P(t) = (1 - t)(1 - qt)Z(C, t)$, which satisfies the functional equation

$$hP(1/qt) = q^{-g}t^{-2g}P(t). \tag{2.8.1}$$

From § 2.7, P is a rational function, and is therefore meromorphic in the whole complex plane. By construction, we have

$$\frac{d \log P(t)}{dt} = -\sum_{n=1}^{\infty} a_n t^{n-1}.$$

From (2.4.1), this series converges for $|t| < q^{-1/2}$, so that P has neither zeros nor poles in this domain. But the functional equation shows that P doesn't have any zeros or poles in the domain $|t| > q^{-1/2}$ either; its zeros and poles must then be concentrated on the circle $|t| = q^{-1/2}$ and $Z(C, t)$ has two simple poles, at $t = 1$ and at $t = 1/q$. By examining (2.7.1), we deduce:

$$\delta = 1; \quad P \text{ is a polynomial.}$$

Lastly, this polynomial has integer coefficients by construction and has degree $2g$ from (2.8.1).

2.9 The Riemann hypothesis: Weil's first proof

This proof relies on the *theory of correspondences* and the *Castelnuovo–Severi inequality*.

2.9.1 Elementary theory of correspondences

(See also [144, I].)

Definition 2.12 Let C, C' be curves (assumed smooth, projective, and geometrically connected) over a field k. An *algebraic correspondence* from C to C' is a divisor on $C \times C'$. This is denoted $\mathrm{Corr}(C, C')$.

Example 2.13 If $f : C \to C'$ is a k-morphism, its graph Γ_f, the image of the closed immersion $C \xrightarrow{\binom{1_C}{f}} C \times C'$, is a correspondence, denoted f_*.

Example 2.14 If $\gamma \in \mathrm{Corr}(C, C')$, then the divisor associated to γ also defines a correspondence from C' to C, which is denoted $^t\gamma \in \mathrm{Corr}(C', C)$ (the transpose correspondence). If γ is of the form Γ_f, we write $^t\gamma = f^*$.

Lemma 2.15 ([Wei2, th. 2 and 5]) *There is a surjection* ("*restriction to the generic point*"[5])

$$\mathrm{Corr}(C, C') \xrightarrow{\rho} \mathrm{Div}(C'_{k(C)}) \to 0$$

whose kernel is given by correspondences of the form $D \times C'$, *where* D *is a divisor of* C. *This surjection has a section* s *sending an irreducible divisor of* $\mathrm{Div}(C'_{k(C)})$ *to its Zariski closure in* $C \times C'$. *Moreover,* ρ *induces an exact sequence*

$$\mathrm{Pic}(C) \xrightarrow{p^*} \mathrm{Pic}(C \times C') \to \mathrm{Pic}(C'_{k(C)}) \to 0 \qquad (2.9.1)$$

where p *is the projection* $C \times C' \to C$.

Proof As Weil stated, this is a consequence of a much more general result: pass to the limit obtained by replacing C by smaller and smaller open subsets in [Ful, prop. 1.8 and his proof]. □

Remark 2.16 If C' admits a 0-cycle z of degree 1 (for example if k is algebraically closed), the sequence (2.9.1) is also left exact, and split by the choice of z.

Definition 2.17 Let C, C', C'' be three curves over the field k, and let $\gamma \in \mathrm{Corr}(C, C')$ and $\gamma' \in \mathrm{Corr}(C', C'')$.

a) We say that γ and γ' are *composable* if the irreducible components of the divisors $\gamma \times C''$ and $C \times \gamma'$ of $C \times C' \times C''$ are pairwise distinct.
b) If γ and γ' are composable, we define their composition $\gamma' \circ \gamma$ to be the divisor

$$(p_{C \times C''}^{C \times C' \times C''})_*(C \times \gamma' \cap \gamma \times C''),$$

where p_* is the direct image of the cycles [Ful, § 1.4] and $\cap$ denotes the intersection product (defined by means of intersection multiplicities)[6].

---

[5] Generic point in the sense of Grothendieck!

[6] Barring too rapid a reading of [Wei1], Weil defines these in the Italian fashion: on a variety which we would now call quasi-projective, by reduction to the diagonal and intersection with a generic linear subspace of complementary dimension; then on an arbitrary variety, by gluing from the quasi-projective case.

If γ and γ' are irreducible, the divisors $\gamma \times C''$ and $C \times \gamma'$ are equal if and only if γ is of the form $C \times P$ and γ' of the form $P \times C''$, where P is a point of C'. This gives

Proposition 2.18 (Weil, [Wei2, th. 6]) *There exists a unique bilinear composition*

$$\circ : \mathrm{Corr}(C', C'') \times \mathrm{Corr}(C, C') \to \mathrm{Corr}(C, C'')$$

such that

(i) *If $\gamma \in \mathrm{Corr}(C, C')$ and $\gamma' \in \mathrm{Corr}(C', C'')$ are composable, their composition is given by the rule in Definition 2.17 b).*
(ii) $(P \times C'') \circ (C \times P) = 0$ *for all* $P \in C'_{(0)}$.

This law is associative (when we use four curves) and the diagonal correspondence $\Delta_C \subset C \times C$ is a left and right identity element. If $f : C \to C'$ and $f' : C' \to C''$ are morphisms, we have

$$(f' \circ f)_* = f'_* \circ f_*.$$

Finally, for all correspondences $\gamma \in \mathrm{Corr}(C, C')$ and $\gamma' \in \mathrm{Corr}(C', C'')$, we have

$${}^t(\gamma' \circ \gamma) = {}^t\gamma \circ {}^t\gamma'.$$

Proof Existence and uniqueness are clear; the formal properties follow from those of the intersection product (cf. [Ful, prop. 16.1.1]). □

2.9.2 Indices

Definition 2.19 (cf. [Ful, ex. 16.1.4]) Let $\gamma \in \mathrm{Corr}(C, C')$ be a correspondence. We associate to it integers $d(\gamma)$, $d'(\gamma)$ called *indices* (or *degrees*) satisfying $(p_C^{C \times C'})_* \gamma = d(\gamma)[C]$ and $(p_{C'}^{C \times C'})_* \gamma = d'(\gamma)[C']$, respectively.

If γ is irreducible, then

$$d(\gamma) = \begin{cases} [k(\gamma) : k(C)] & \text{if } \gamma \text{ is dominant on } C, \\ 0 & \text{otherwise} \end{cases} \qquad (2.9.2)$$

and similarly for $d'(\gamma)$.

Remark 2.20 We have $d(\gamma) = \deg \rho(\gamma)$, where ρ is the homomorphism of Lemma 2.15.

Lemma 2.21 *We have identities*

$$d(^t\gamma) = d'(\gamma), \quad d(\gamma' \circ \gamma) = d(\gamma')d(\gamma).$$ □

Proposition 2.22 (Weil, [Wei2, prop. 4]) *Let $\gamma \in \mathrm{Corr}(C, C')$ be a positive correspondence, without components of the form $P \times C'$ where P is a closed point of C, and such that $d(\gamma) = 1$. Then*

$$\gamma \circ {}^t\gamma = d'(\gamma)\Delta_{C'}.$$

Equivalently (Exercise 2.23 (b)), if $f : C \to C'$ is a non-constant morphism, then $f_* \circ f^* = \deg(f)\Delta_{C'}$.

Exercise 2.23

(a) Prove the formula in Proposition 2.22.
(b) Show that γ satisfies the hypothesis of this proposition if and only if it is of the form f_*, where f is a morphism from C to C'. (Use the fact that a rational map from C to C' is everywhere defined, by the valuative criterion for properness.)

2.9.3 The trace

Definition 2.24 Consider $\gamma_1, \gamma_2 \in \mathrm{Corr}(C, C')$ as divisors. We denote their *intersection number* by $I(\gamma_1, \gamma_2) \in \mathbf{Z}$.

In Weil, this number is defined as follows [Wei2, p. 30]: if γ, γ' are distinct reduced and irreducible curves, their intersection $\gamma \cap \gamma'$ is a zero-cycle and $I(\gamma, \gamma') = \deg(\gamma \cap \gamma')$. This extends to divisors without common components by linearity. If $\gamma = \gamma'$, then there exists a rational function f such that $v_\gamma(f) = 1$, and we define $I(\gamma, \gamma)$ as $I(\gamma, \gamma - (f))$. This definition works, since γ and (f) have no common component, so $I(\gamma, (f)) = 0$ (which can be seen by restricting f to the normalisation of the components of γ). Then Weil extends by bilinearity.

Exercise 2.25 ([Wei2, th. 3]) Let $\gamma \in \mathrm{Corr}(C, C')$. Show that

$$I(\gamma, D \times C') = d(\gamma) \deg(D)$$

for all divisors D on C.

Proposition 2.26 ("Hurwitz formula", [Wei2, prop. 2]) *We have the identity*

$$I(\gamma_1, \gamma_2) = I(^t\gamma_2 \circ \gamma_1, \Delta_C).$$

Proof This is a combinatorial exercise, cf. [Ful, ex. 16.1.3] for a more precise statement (and a proof). □

Definition 2.27 Let C be a curve and let $\gamma \in \mathrm{Corr}(C, C)$. We define the *trace* of γ to be

$$\sigma(\gamma) = d(\gamma) + d'(\gamma) - I(\gamma, \Delta_C).$$

(For a motivic interpretation, see Exercise 6.45.)

Definition 2.28 Two correspondences $\gamma_1, \gamma_2 \in \mathrm{Corr}(C, C')$ are *equivalent* (denoted $\gamma_1 \equiv \gamma_2$) when there exists a rational function f and divisors D on C and D' on C' such that

$$\gamma_1 - \gamma_2 = (f) + D \times C' + C \times D'.$$

This is the *three-line equivalence* of Weil (in J.P. Murre's terminology). We write $\mathrm{Corr}_\equiv(C, C') = \mathrm{Corr}(C, C')/ \equiv$: this is the group of *correspondence classes* (in the extended sense).

The correspondences that are $\equiv 0$ are called *of valence* 0 in the terminology of the Italian school of geometers.

Exercise 2.29 Let K be an extension of k. Show that $\sigma(\gamma) = \sigma(\gamma_K)$ for all $\gamma \in \mathrm{Corr}(C, C)$, where γ_K is the image of γ in $\mathrm{Corr}(C_K, C_K)$ under extension of scalars.

Proposition 2.30

(1) **[Wei2, p. 38]**. *The set of $\gamma \equiv 0$ is a two-sided ideal in $\mathrm{Corr}(C, C)$. More generally, the composition of Proposition 2.18 induces a composition*

$$\circ : \mathrm{Corr}_\equiv(C', C'') \times \mathrm{Corr}_\equiv(C, C') \to \mathrm{Corr}_\equiv(C, C'').$$

(2) **[Wei2, th. 7]**. *The trace σ induces a function*

$$\sigma : \mathrm{Corr}_\equiv(C, C) \to \mathbf{Z}$$

which satisfies the identity $\sigma(^t\gamma) = \sigma(\gamma)$.

(3) *(ibid.). If $\gamma \in \mathrm{Corr}_\equiv(C, C')$ and $\gamma' \in \mathrm{Corr}_\equiv(C', C)$, we have the relation*

$$\sigma(\gamma' \circ \gamma) = \sigma(\gamma \circ \gamma').$$

Proposition 2.31 *The exact sequence* (2.9.1) *of Proposition 2.15 induces isomorphisms*

$$\mathrm{Corr}_\equiv(C, C') \xrightarrow{\sim} \mathrm{Pic}(C'_{k(C)})/p'^* \mathrm{Pic}(C') \xleftarrow{\sim} \mathrm{Pic}^0(C'_{k(C)})/p'^* \mathrm{Pic}^0(C').$$

Proof By definition of $\equiv$, we have an isomorphism

$$\mathrm{Pic}(C \times C')/p^* \mathrm{Pic}(C) + p'^* \mathrm{Pic}(C') \xrightarrow{\sim} \mathrm{Corr}_\equiv(C, C'),$$

where $p : C \times C' \to C$ and $p' : C \times C' \to C'$ are the projections. The isomorphism on the left follows immediately, and the one on the right follows from a short diagram chase. $\square$

Exercise 2.32 (Swan, [123]) Let $K = k(C)$ and $K' = k(C')$.

(a) Show that $R = K \otimes_k K'$ is a Dedekind ring.
(b) Show that the natural homomorphism $\mathrm{Pic}(C \times C') \to \mathrm{Pic}(\mathrm{Spec}\, R)$ induces an isomorphism

$$\mathrm{Corr}_\equiv(C, C') \xrightarrow{\sim} \mathrm{Pic}(\mathrm{Spec}\, R).$$

Exercise 2.33

(a) Let X be a smooth variety over k, and let K/k be a Galois extension. Show that $\mathrm{Pic}(X) \to \mathrm{Pic}(X_K)$ is injective. (Use Hilbert's Theorem 90.)
(b) Suppose that C has genus 0. Show that

$$\mathrm{Corr}_\equiv(C, C) = \begin{cases} 0 & \text{if } C \text{ has a rational point,} \\ \mathbf{Z}/2 & \text{otherwise.} \end{cases}$$

(In the first case, $C \simeq \mathbf{P}^1$, cf. Exercise 2.9 (b); show that the class of the diagonal Δ_C is then equal to $C_1 + C_2$, where C_1, C_2 are the two generators. In the second case, let $c \in C_{(0)}$ be a separable point of degree 2, and let $K = k(c)$; by setting $X = C \times C$ and taking (a) into account, show that $\mathrm{Pic}(X) \subset \mathrm{Pic}(X_K)$ contains $2C_1, 2C_2$ and $C_1 + C_2$. Conclude by showing that $C_1 \notin \mathrm{Pic}(X)$; for this you can use the map "restriction to the diagonal" $\mathrm{Pic}(X) \to \mathrm{Pic}(\Delta_C)$.)

Proposition 2.34 (Weil, [Wei2, prop. 3]) *Suppose k is algebraically closed. Let $\gamma \in \mathrm{Corr}_\equiv(C, C')$. There exists a positive correspondence X in the class γ, without any component of the form $D \times C'$ and satisfying $d(X) = g' := g(C')$.*

Proof Let X_1 be a correspondence in the class γ, and set $d = d(X_1)$. Let $\mathfrak{n} \in \mathrm{Div}(C')$ be a divisor of degree $g' - d$ (which exists since k is algebraically closed). We then apply Riemann–Roch (2.6.1) to the divisor of degree g'

$$\mathfrak{m}_1 = \rho(X_1) + \mathfrak{n}_{k(C)} \in \mathrm{Div}(C'_{k(C)}),$$

where ρ is the projection from Lemma 2.15. Then there exists $\varphi \in k(C)^*$ such that $\mathfrak{m} = \mathfrak{m}_1 + (\varphi)$ is positive. We get X by applying the section s of Lemma 2.15 to $\mathfrak{m}$. $\square$

Remark 2.35 In Proposition 2.34, the condition on X tells us that the projection onto C of each of its irreducible components is finite and surjective; in other words, γ is a *finite correspondence* in the sense of Voevodsky [135, § 2.2]. The statement is then similar to a theorem of Friedlander–Voevodsky (loc. cit., proof of Prop. 2.1.4).

2.9.4 Positivity of the trace form

Proposition 2.36 (Weil, [Wei2, th. 8]) *If the curve C has genus g, then $\sigma(\Delta_C) = 2g$.*

Sketch of proof As $d(\Delta_C) = d'(\Delta_C) = 1$, the statement is equivalent to the equality $I(\Delta_C, \Delta_C) = 2 - 2g$. This follows from the Riemann–Roch theorem (2.6.1) because if f is a rational function on $C \times C$ satisfying $v_{\Delta_C}(f) = 1$, then $(p_C^{C \times C})_*(((f) - \Delta_C) \cap \Delta_C))$ (the first projection) is a canonical divisor [Wei2, p. 15]. $\square$

Theorem 2.37 (Weil, [Wei2, th. 10]) *Suppose that k is algebraically closed, and let $\gamma \in \mathrm{Corr}_{\equiv}(C, C') - \{0\}$. Then we have $\sigma(\gamma \circ {}^t\gamma) > 0$.*

(The hypothesis that k is algebraically closed is necessary, cf. Exercises 2.29 and 2.33.)

If $k = \mathbf{C}$, this inequality is due to Guido Castelnuovo [21]; see also Francesco Severi [117].

Corollary 2.38 *Suppose that k is algebraically closed.*

(1) *The map*

$$(\gamma_1, \gamma_2) \mapsto \sigma(\gamma_2 \circ {}^t\gamma_1)$$

from $\mathrm{Corr}_{\equiv}(C, C') \times \mathrm{Corr}_{\equiv}(C, C')$ to $\mathbf{Z}$ is a positive-definite symmetric bilinear form.

(2) *For all correspondence classes $\gamma \in \mathrm{Corr}_{\equiv}(C, C)$, we have the inequality*

$$\sigma(\gamma)^2 \le 2g\sigma(\gamma \circ {}^t\gamma),$$

with equality if and only if γ and Δ_C are linearly dependent.

Proof (1) The symmetry results from Proposition 2.30 and the positive-definiteness from Theorem 2.37.

(2) The quadratic form

$$(a, b) \mapsto \sigma \left({}^t(a\Delta_C + b\gamma)(a\Delta_C + b\gamma)\right) = 2ga^2 + 2\sigma(\gamma)ab + \sigma(\gamma \circ {}^t\gamma)b^2$$

(cf. Proposition 2.36 and Proposition 2.30) is positive, thus its discriminant is either negative or zero; it is zero if and only if the form is not defined, which happens precisely when γ and Δ_C are linearly dependent (in $\mathrm{Corr}_{\equiv}(C, C)$).
□

We quote Weil (Œuvres, comments on [139], p. 557):

I finally realized that it wasn't so much my lemma that I needed; everything relied on the properties of the trace σ and primarily on the positivity of $\sigma(\gamma\gamma')$. Now this is a question of pure algebraic geometry provided we introduce σ not as the trace of a matrix, but as the formula of the fixed points which thus becomes its definition. It is then easy to check that σ has the formal properties of a trace on the ring of correspondence classes.[7]

See § 3.2.2 for a description of σ as an actual trace.

2.9.5 Sketch of the proof of Theorem 2.37

This occupies pages 47–54 of [Wei2]: to explain it in detail in the language of schemes would be beyond the scope of this book. Instead we give its general idea:

If $g = 0$, the statement is vacuous (Exercise 2.33). Suppose that $g = 1$ (this case is due to Hasse [49]): by Proposition 2.34 we may represent γ as a correspondence X such that $d(X) = 1$. By using Proposition 2.36 we find

$$\sigma(\gamma \circ {}^t\gamma) = d'(\gamma)\sigma(\Delta_C) = 2d'(\gamma).$$

From (2.9.2) we have $d'(\gamma) \geq 0$, with equality if and only if γ has support in $C \times F$, where F is an appropriate closed proper subset of C'. But if this is the case then we have $\gamma = 0$ in $\mathrm{Corr}_{\equiv}(C, C')$.

In the case $g > 1$, Weil's reasoning is much more delicate; variants have subsequently been given by Igusa [54], Roquette [97], Eichler [Eic], Swan

---

[7] *Je finis par comprendre que ce n'était pas tant mon lemme dont j'avais besoin; tout tenait aux propriétés de la trace σ et principalement à la positivité de $\sigma(\gamma\gamma')$. Or c'est là une question de pure géométrie algébrique pourvu qu'on introduise σ, non comme trace d'une matrice, mais par la formule des points fixes qui en devient ainsi la définition. Il est facile alors de vérifier que σ a les propriétés formelles d'une trace sur l'anneau des classes de correspondances.*

[123] and Kani [67]. Everywhere, the first step is to strengthen Proposition 2.34 to a statement in which the correspondence X is restricted to "non-special divisors" on C'. Then the methods diverge: Weil, Roquette, Eichler and Swan use an auxiliary rational function φ on C, constructed by using a determinant. A thorough examination of the divisor of φ leads to a stronger inequality than that of Theorem 2.37. Igusa, meanwhile, uses a formula from intersection theory due to Schubert in the case $k = \mathbf{C}$, which he generalises to an arbitrary algebraically closed field. Kani's approach [67] is different again, using the language of schemes, as well as precise references to early work by Castelnuovo and Severi.

2.9.6 Mattuck–Tate and Grothendieck

Let X be a smooth projective k-variety. A divisor $D \in \mathrm{Div}(X)$ is said to be *numerically equivalent to zero* if $I(D, \Gamma) = 0$ for all curves $\Gamma \subset X$ that intersect D properly. (This is a special case of a more general notion, cf. [Ful, def. 19.1] and Examples 6.2.) If D is rationally equivalent to zero, then it is numerically equivalent to zero. We denote the subgroup of divisor classes numerically equivalent to zero by $\mathrm{Pic}^\tau(X) \subset \mathrm{Pic}(X)$, and write $N(X) = \mathrm{Pic}(X)/\mathrm{Pic}^\tau(X)$.

Since the trace σ of Definition 2.27 is expressed in terms of intersection numbers (Exercise 2.25), we have $\sigma(D) = 0$ for all $D \in \mathrm{Pic}^\tau(C \times C')$. In this way, σ factorises as a homomorphism

$$\bar{\sigma} : \overline{\mathrm{Corr}}_{\equiv}(C, C') \to \mathbf{Z},$$

where $\overline{\mathrm{Corr}}_{\equiv}(C, C')$ is the quotient of $\mathrm{Corr}_{\equiv}(C, C')$ by the subgroup of divisor classes numerically equivalent to zero.

In [86], Arthur Mattuck and John Tate prove a weakened version of Theorem 2.37 which is sufficient to prove the Riemann hypothesis.

Theorem 2.39 *We have*

$$\bar{\sigma}(\gamma \circ {}^t\gamma) > 0 \quad if \quad \gamma \in \overline{\mathrm{Corr}}_{\equiv}(C, C') - \{0\}. \tag{2.9.3}$$

Mattuck and Tate deduced Theorem 2.39 from the Riemann–Roch theorem for the surface $C \times C'$. Alexander Grothendieck then remarked in [42] (which is written in Weil's style!) that this follows directly from the *Hodge index theorem for a surface*, due to Beniamino Segre [106] and Jacob Bronowski [16], and which he rediscovered at the time. Here it is.

Theorem 2.40 *Let S be a smooth projective surface over k, and let H be an ample divisor on S. Denote the intersection product of divisors by $\langle , \rangle$. Let D*

be a divisor on S such that $\langle D, H \rangle = 0$. *Then* $\langle D, D \rangle \leq 0$, *with equality if and only if D is numerically equivalent to zero.*

Proof See [Har2, ch. V, th. 1.9] or [Ful, ex. 15.2.4]. As in Mattuck–Tate, the proof uses the Riemann–Roch theorem for S. □

Proof of Theorem 2.39. We give the version of Grothendieck's argument offered by Fulton in [Ful, ex. 16.1.10 (a)]. By Proposition 2.26, we must show that if $\gamma \in \mathrm{Corr}(C, C)$ is not $\equiv 0$, then

$$I(\gamma, \gamma) < d({}^t\gamma \circ \gamma) + d'({}^t\gamma \circ \gamma).$$

In view of Lemma 2.21, this inequality is equivalent to

$$I(\gamma, \gamma) < 2d(\gamma)d'(\gamma),$$

which follows from Theorem 2.40 for the surface $S = C \times C'$, applied to the divisor $D = \gamma - d(\gamma)(C \times P') - d'(\gamma)(P \times C')$ and to the ample divisor $H = C \times P' + P \times C'$, where P (resp. P') is a point of C (resp. of C'), cf. Exercise 2.25. □

2.9.7 Comparison between Theorems 2.37 and 2.39

Putting these together, we get

Corollary 2.41 *We have an isomorphism*

$$\mathrm{Corr}_{\equiv}(C', C) \xrightarrow{\sim} \overline{\mathrm{Corr}}_{\equiv}(C', C).$$ □

Proposition 2.42 *Corollary 2.41 is equivalent to the following statement: the projections* $p : C \times C' \to C$ *and* $p' : C \times C' \to C'$ *induce an isomorphism*

$$\mathrm{Pic}^\tau(C) \oplus \mathrm{Pic}^\tau(C') \xrightarrow{\sim} \mathrm{Pic}^\tau(C \times C'). \tag{2.9.4}$$

Proof This follows easily from the snake lemma, using auxiliary rational points on C and C'. □

For any smooth projective k-variety X, we write $\mathrm{NS}(X)$ for the *Néron–Severi group* of X: the group of divisor classes modulo *algebraic equivalence* [Ful, def. 10.3]. We also write

$$\mathrm{Pic}^0(X) = \mathrm{Ker}(\mathrm{Pic}(X) \to \mathrm{NS}(X));$$

if X is a curve, this group coincides with the one defined in § 2.5. We have $\mathrm{Pic}^0(X) \subset \mathrm{Pic}^\tau(X)$, and a theorem of Teruhisa Matsusaka [85] tells us that $\mathrm{Ker}(\mathrm{NS}(X) \to N(X)) = \mathrm{Pic}^\tau(X)/\mathrm{Pic}^0(X)$ is a finite group for all smooth projective k-varieties X. Starting from this, the isomorphism (2.9.4) is attributed by Grothendieck in [42, (1.9)] to unpublished work of Serre: it holds more generally for any two smooth projective k-varieties.

Remark 2.43 Since $\mathrm{Pic}^0(C) = \mathrm{Pic}^\tau(C)$ and $\mathrm{Pic}^0(C') = \mathrm{Pic}^\tau(C')$, we deduce from (2.9.4) that $\mathrm{Pic}^0(C \times C') = \mathrm{Pic}^\tau(C \times C')$. Therefore, the Néron–Severi group is *torsion-free* for products of two curves. In fact, it is torsion-free for arbitrary products of curves, as one sees by using (2.9.4) repeatedly.

Thus we can recover Theorem 2.37 from Theorem 2.39 and the isomorphism (2.9.4); but this is not how Weil proceeds! We will see another way of recovering Theorem 2.37 in § 3.2.2, using the theory of abelian varieties.

2.9.8 Case of the graph of an endomorphism

Let $f : C \to C$ be an endomorphism, with graph Γ_f. By applying Corollary 2.38[8] to Γ_f, taking into account Proposition 2.22, and by reusing Proposition 2.36, we obtain the following result.

Theorem 2.44 *We have the inequality*

$$|\sigma(f_*)| \le 2g\sqrt{\deg(f)}.$$

If Γ_f is transverse to Δ_C, this inequality becomes

$$|1 + \deg(f) - N(f)| \le 2g\sqrt{\deg(f)}$$

where $N(f)$ is the number of fixed points of f.

Proof Since $d(f_*) = 1$ and $d'(f_*) = \deg(f)$, in order to deduce the second inequality from the first it suffices to note that $I(f_*, \Delta_C) = N(f)$ by the assumption of transversality. $\square$

2.9.9 End of the proof

To prove (2.4.1), Weil extends scalars from $\mathbf{F}_q$ to $\bar{\mathbf{F}}_q$ and applies Theorem 2.44 to the powers of the *Frobenius correspondence* π_C: the graph of the Frobenius morphism

$$F_C : C \to C$$

defined (in Grothendieckian terms) to be the identity on the points of C and by the map $f \mapsto f^q$ on the sheaf of rings $\mathcal{O}_C$. As F_C is radical, its graph is

---

[8] The version for $\overline{\mathrm{Corr}}_\equiv(C', C)$, which follows from Theorem 2.39, suffices.

indeed transverse to Δ_C as well as to that of F_C^n for all $n > 0$: see [Wei2, proof of th. 13]. It then suffices to note that the set of fixed points of F_C^n is $C(\mathbf{F}_{q^n})$.

2.10 First applications

Corollary 2.45

(1) *Let C be a smooth projective curve of genus g that is geometrically connected over $\mathbf{F}_q$. Then*

$$1 + q - 2g\sqrt{q} \leq |C(\mathbf{F}_q)| \leq 1 + q + 2g\sqrt{q}.$$

(2) *Let C be a smooth projective curve of genus g that is geometrically irreducible over a number field k. If v is a finite place of k, of good reduction for C (§5.2), then C has a rational point on k_v whenever $N(v) \geq 4g^2$.*

Proof (1) is immediate from (2.4.1). This implies $X(\mathbf{F}_q) \neq \varnothing$ whenever $1 + q - 2g\sqrt{q} > 0$. From this it follows that

$$\sqrt{q} > g + \sqrt{g^2 - 1}$$

which holds whenever $\sqrt{q} \geq 2g$ or $q \geq 4g^2$. For (2), we deduce that the special fibre at v has a rational point whenever $N(v) > 4g^2$, so that C also has a rational point over k_v, by Hensel's lemma. $\qquad\qquad\square$

Example 2.46 If C is a curve of genus 1 over $\mathbf{Q}$ with good reduction at p, then C has a rational point over $\mathbf{Q}_p$ whenever $p \geq 5$.

Corollary 2.47 *Let C be a (possibly open and singular) curve over $\mathbf{F}_q$ that is geometrically integral. Then $\zeta(C, s)$ is a rational function in q^{-s}, its poles are simple and located at $s = 1 + \frac{2i\pi r}{\log q}$ for $r \in \mathbf{Z}$, and it is non-vanishing for $\Re(s) > \frac{1}{2}$.*

Proof Suppose first that C is proper, and let $\tilde{C}$ be its normalisation. The projection $f : \tilde{C} \to C$ is an isomorphism away from a finite number of points $c_1, \ldots, c_n \in C$. We have

$$Z(c_i, t) = (1 - t^{d_i})^{-1}$$

where $d_i = \deg(c_i)$ and

$$Z(f^{-1}(c_i), t) = \prod_{c \in f^{-1}(c_i)} (1 - t^{d_c})^{-1},$$

where $d_c = \deg(c)$. As $d_i \mid d_c$ for all c, we have $(1 - t^{d_i}) \mid \prod(1 - t^{d_c})$. We then conclude that

$$Z(C, t) = Z(\tilde{C}, t)Q(t) \tag{2.10.1}$$

where Q is a polynomial whose zeros are roots of unity.

If C is arbitrary, we consider its completion $\bar{C}$: if $\tilde{C}$ is the normalisation of $\bar{C}$, we see that (2.10.1) remains true, with the same property for Q. The conclusion then follows from Theorem 2.7. □

2.11 The Lang–Weil theorems

The following corollary is due to Lang–Weil [76] and to Nisnevič [88] independently (the latter gave a slightly weaker bound).

Corollary 2.48 *Let X be a geometrically irreducible subvariety of $\mathbf{P}^N_{\mathbf{F}_q}$ of dimension n and degree d. Then*

$$\left| |X(\mathbf{F}_q)| - q^n \right| \leq (d-1)(d-2)q^{n-1/2} + Aq^{n-1}$$

where A is a constant depending only on n, d, and N.
In particular, X admits a zero-cycle of degree 1.

Sketch of proof Lang and Weil proceed by induction on n. The case $n = 1$ is a variation on the proof of Corollary 2.47, using the identity

$$\frac{(d-1)(d-2)}{2} = g + \sum_P \delta_P,$$

where g is the genus of $\tilde{X}$, the desingularisation of X, and δ_P is the length of the $\mathcal{O}_{X,P}$-module $\mathcal{O}_{\tilde{X},P}/\mathcal{O}_{X,P}$ for all $P \in X$ [Har2, I, ex. 7.2; IV, ex. 1.8].

For $n > 1$, we may suppose without loss of generality that X is not contained in any hyperplane H of $\mathbf{P}^N$; all intersections $H \cap X$ must then be *proper*. The locus of those H such that $H \cap X$ is reducible is a closed algebraic subset Z of the dual projective space $\mathbf{P}^N$ [144, lem. 4]. We easily see that Z is contained in a hypersurface, which bounds the number of $\mathbf{F}_q$-points by Bq^{N-1} where B is a constant depending on (n, d, N). An elementary counting argument allows one to conclude the proof. □

Remark 2.49 It is not necessary to apply the estimate from Corollary 2.48 in order to prove the existence of a zero-cycle of degree 1 on X: it suffices to apply the result of § 2.8 to the normalisation of a curve drawn on X.

Corollary 2.50 *Let K be a function field in n variables over $\mathbf{F}_q$, which we assume to be algebraically closed in K. Then there exists a constant γ such that, for all $\mathbf{F}_q$-models V of K and all $r \geq 1$, we have the inequality*

$$\left| |V(\mathbf{F}_{q^r})| - q^{rn} \right| \leq \gamma q^{r(n-1/2)} + Bq^{r(n-1)},$$

where the constant B depends on V. If K admits a projective model of degree d, then we may take $\gamma \leq (d-1)(d-2)$.

Proof We argue by induction on n, the case $n = 0$ being trivial (with $\gamma = B = 0$). For $n > 0$, this shows that the statement is invariant under passage to a nonempty open subset of V; we may then suppose that V is affine, hence projective: this case is dealt with in Corollary 2.48. □

Corollary 2.51 *If V' is another model of K, then the function $\frac{\zeta(V,s)}{\zeta(V',s)}$ converges absolutely for $\Re(s) > n - 1$; the divisor of the zeros and poles of $\zeta(V, s)$ in this region is therefore a birational invariant of K.* □

The following corollary is due to Lang–Weil [76] when V is as above, and is generalised by Serre in [112].

Corollary 2.52

(1) *Let V be a **Z**-scheme of finite type of dimension n. Then $\zeta(V, s)$ extends analytically to the half-plane $\Re(s) > n - \frac{1}{2}$.*
(2) *Suppose V is integral with function field E.*
   (i) *If char $E = 0$, then $\zeta(V, s)$ has a unique simple pole in this domain, at $s = n$.*
   (ii) *If char $E = p$, let $\mathbf{F}_q$ be its field of constants (the algebraic closure of $\mathbf{F}_p$ in E). Then the only poles of $\zeta(V, s)$ are simple and lie at the points*

$$s = n + \frac{2i\pi r}{\log q}, \quad r \in \mathbf{Z}.$$

Proof Serre suggested a proof using a fibration of curves; we will proceed differently, using Corollary 2.50. We first go from (1) to (2) by reasoning by induction on n as in point (1) of the proof of Theorem 2.5. In case (ii) of (2), the statement follows immediately from Corollary 2.50 by using the formula from Proposition 2.3 (3): more precisely, the latter shows that, as in [76, p. 826], the series $Z(V, t)(1 - q^n t)$ converges absolutely for $|t| < q^{-(n-\frac{1}{2})}$.

In case (i), let K be the algebraic closure of $\mathbf{Q}$ in E: from [EGA4$_3$, Prop. 9.7.8][9], all but finitely many of the fibres of the morphism $V \to \operatorname{Spec} O_K$ are geometrically irreducible of dimension $n - 1$. We then deduce from Corollary 2.50 and the abscissa of absolute convergence of $\zeta(O_K, s)$ (§ 1.10) that $\zeta(V, s)\zeta(O_K, s - n + 1)$ converges absolutely for $\Re(s) > n - \frac{1}{2}$, which gives the conclusion, again using 1.10. □

---

[9] I thank Colliot-Thélène for this reference.

3

The Weil conjectures

3.1 From curves to abelian varieties

Let us briefly recall, without proofs, the most important points of the theory of abelian varieties (references: [Wei3], [Lan1], [Mum], [83, 84]). For a historical overview, see Milne's bibliographic notes at the end of [84] and Weil's comments in his scientific works.

3.1.1 Generalities

Definition 3.1 Let k be a field. An *abelian variety* over k is a proper, integral, algebraic k-group.

Note that integral $\Rightarrow$ geometrically integral (since A admits a rational point, the identity element) $\Rightarrow$ smooth. (In fact, we may reduce to k algebraically closed; the open set of smooth points in A is therefore nonempty, and we use translations to cover A.)

Theorem 3.2

(1) *An abelian variety is projective and its group law is commutative.*
(2) *Let $f : A \rightarrow B$ be a k-morphism of abelian varieties. For all points $b \in B(k)$, denote by t_b the (left or right)-translation by b. Then $t_{-f(0)} \circ f : a \mapsto f(a) - f(0)$ is a homomorphism.*

Definition 3.3 Let A, B be abelian varieties over k. An *isogeny* from A to B is a surjective (i.e. faithfully flat) homomorphism $f : A \rightarrow B$ with finite kernel.

Theorem 3.4 *Let A be an abelian variety of dimension g. Let l be a prime number distinct from $\mathrm{char}\, k$, and let $\bar{k}$ be an algebraic closure of k. Then $A(\bar{k})$ is divisible, and the group $_{l^n}A(\bar{k})$ is (non-canonically) isomorphic to $(\mathbf{Z}/l^n)^{2g}$*

for all $n \geq 1$. Then the Tate module

$$T_l(A) = \varprojlim {}_{l^n} A(\bar{k})$$

is a free $\mathbf{Z}_l$-module of rank $2g$.[1]

Theorem 3.5 *Let A, B be two abelian varieties over k. Then $\mathrm{Hom}_k(A, B)$ is a free $\mathbf{Z}$-module of finite type, and the homomorphism*

$$\mathrm{Hom}_k(A, B) \otimes \mathbf{Z}_l \to \mathrm{Hom}_{\mathbf{Z}_l}(T_l(A), T_l(B))$$

is injective for all $l \neq \mathrm{char}\, k$. Furthermore, the algebra

$$\mathrm{End}^0(A) = \mathrm{End}(A) \otimes \mathbf{Q}$$

is semisimple.

Corollary 3.6 *Any surjective endomorphism of an abelian variety is an isogeny.*

Corollary 3.7 (Poincaré's complete reducibility theorem) *Let A be an abelian k-variety, and let B be an abelian sub-k-variety of A. Then there exists an abelian sub-k-variety C of A such that $B + C = A$ and $B \cap C$ is finite.*

Over $k = \mathbf{C}$, all these results are easily obtained from the transcendental description of an abelian variety, as the quotient of $\mathbf{C}^g$ by a lattice with Riemann forms. Over an arbitrary field (especially of characteristic > 0), these are considerably harder, and are due to Weil [Wei3] (except the one of Footnote[1]; projectivity is also due independently to Matsusaka and to Barsotti).

For the sake of completeness, we quote

Theorem 3.8 (Tate, Zarhin, Mori, Faltings) *Let k be a field of finite type (over its prime subfield), and let A, B be two abelian k-varieties. Let k_s denote the separable closure of k and $G = \mathrm{Gal}(k_s/k)$ the absolute Galois group of k. Finally, let l be a prime number distinct from $\mathrm{char}\, k$. Then the homomorphism*

$$\mathrm{Hom}_k(A, B) \otimes \mathbf{Z}_l \to \mathrm{Hom}_{\mathbf{Z}_l}(T_l(A), T_l(B))^G$$

is an isomorphism.

When k is finite, this gives a complete classification of abelian varieties over k, see [127] or [Mum, App. 2, th. 2 and 3]. See also § 6.14.

---

[1] If $l = p = \mathrm{char}\, k$, then ${}_{p^n} A(\bar{k}) \simeq (\mathbf{Z}/p^n)^r$ with $0 \leq r \leq g$. Furthermore, ${}_{p^n} A$ has the structure of a *finite group k-scheme* of $\mathbf{Z}/p^n$-rank $2g$: see [Mum, end of § 6 and § 15, "The p-rank"].

3.1.2 The characteristic polynomial of an endomorphism

Definition 3.9 Let $f \in \text{End}(A)$. We write $\deg(f)$ for the degree of f if f is surjective, and set $\deg(f) = 0$ otherwise.

Example 3.10 $\deg(n 1_A) = n^{2g}$.

Theorem 3.11 *Let A be an abelian variety of dimension g, and let $f \in \text{End}(A)$. Then*

$$\det T_l(f) = \deg(f)$$

for all $l \neq \text{char } k$. Thus,

$$\deg(n 1_A - f) = P(n) \quad \text{for all } n \in \mathbf{Z},$$

where P is the characteristic polynomial of $T_l(f)$.
In particular, P has integer coefficients, degree $2g$, and is independent of l.

Proof See [Mum, § 19, th. 4] or [83, prop. 12.9]. ☐

3.1.3 Picard varieties; duality

Theorem 3.12 *Let X be a smooth projective k-variety. If k is algebraically closed, the Picard group $\text{Pic}(X)$ can be expressed as an extension*

$$0 \to \text{Pic}^0(X) \to \text{Pic}(X) \to \text{NS}(X) \to 0,$$

where $\text{NS}(X)$, the Néron–Severi group of X, is an abelian group of finite type, and $\text{Pic}^0(X) = \text{Pic}^0_{X/k}(k)$ is the group of k-points of an abelian variety, the Picard variety of X.

If k is arbitrary, $\text{Pic}^0_{X_{\bar{k}}/\bar{k}}$ is defined over k (notation: $\text{Pic}^0_{X/k}$).

If X is a curve of genus g, then $\text{Pic}^0_{X/k}$ is of dimension g: we call it the Jacobian of X and denote it (here) by $J(X)$.

Definition 3.13 Let A be an abelian variety; we write A^* for its Picard variety[2], which we call the *dual variety* of A.

A partial justification for this terminology is that every homomorphism $f : A \to B$ induces a dual homomorphism $f^* : B^* \to A^*$.

Theorem 3.14 (biduality) *We have $A^{**} = A$. If X is a curve, $J(X)$ is canonically isomorphic to its dual (see, for example, 3.30).*

---

[2] or sometimes A', $\hat{A}$, $A^\vee$, ...

Corollary 3.15 *If* $f : A \to B$ *is an isogeny, then* f^* *is also an isogeny and* $\deg(f^*) = \deg(f)$.

More precisely, the group schemes Ker f and Ker f^* are dual. The duality in question is Cartier duality for finite group schemes: the dual of such a group scheme G is $G^* = \mathrm{Hom}(G, \mathbf{G}_m)$. See [Mum, § 15, th. 1] or [83, § 11] for the proof.

The best explanation of the duality of abelian varieties is *via* the theory of bi-extensions (Mumford, Grothendieck [SGA7, exp. VII-VIII]), which leads to the notion of 1-motive (Deligne [27, § 10]).

Definition 3.16 Let l be a prime number distinct from char k. The pairing

$$T_l(A) \times T_l(A^*) \to \mathbf{Z}_l(1) := \varprojlim \mu_{l^n}$$

obtained by applying Corollary 3.15 to the groups $_{l^n}A = \mathrm{Ker}(l^n)$ and their duals is called the *Weil pairing*.

3.1.4 Morphisms to an abelian variety

Theorem 3.17 *Let* X *be a smooth k-variety, U be a dense open of X and* $f : U \to A$ *be a morphism from U to an abelian variety. Then f extends (uniquely) to X.*

Proof See [Wei3, n° 15, th. 6] or [83, th. 3.1]. □

3.1.5 The Albanese variety

The Albanese variety of a smooth projective variety X/k is an abelian variety that is, roughly, universal for morphisms X to an abelian variety. Unfortunately the universal property just stated is incorrect, and the correct statement is somewhat less elegant.

In the case in which X has a rational point x_0, the correct statement is simple: we use the k-morphisms $f : X \to A$ such that $f(x_0) = 0$. What is the correct statement when X does not necessarily have a rational point?

Here are two equivalent such statements:

Definition 3.18 Let X be a smooth projective k-variety, and let A be an abelian k-variety. A morphism $f : X \times_k X \to A$ is called *admissible* if $f(\Delta_X) = 0$.

(Given Theorem 3.17, this reduces to stating the universal problem "over $K = k(X)$" with the generic point of X now K-rational.)

Theorem 3.19 *The universal problem given by Definition* 3.18 *has a solution* $\text{Alb}(X)$.

To state the second universal property, we need the notion of a *torsor* (or *principal homogeneous space*) with respect to a group scheme:

Definition 3.20 Let S be a scheme, and let G be a group scheme over S.

a) The *trivial G-torsor* over S is G, equipped with the left action of G.
b) A *G-torsor with base S* is a faithfully flat S-scheme E, equipped with a group action of G (over S)

$$\mu : G \times_S E \to E$$

such that the diagram

$$
\begin{array}{ccc}
G \times_S E & \overset{\mu}{\longrightarrow} & E \\
{\scriptstyle p_2}\downarrow & & \downarrow \\
E & \longrightarrow & S
\end{array}
$$

is cartesian.

(The condition of being "cartesian" means that the morphism $\varphi : G \times_S E \to E \times_S E$ defined by p_2 and μ is an isomorphism: in other words, the pullback of the G-torsor E from S to E is trivial.)

Theorem 3.21 *Let X be a smooth projective k-variety. Then the universal property defined by the triples (A, P, f) where A is an abelian k-variety, P is an A-torsor with base k and $f : X \to P$ is a k-morphism, has a solution* $(\text{Alb}(X), P_X, f_X)$.

Example 3.22 Starting with an abelian variety A and an A-torsor P with base k, we have $\text{Alb}(P) = A$, $P_P = P$ and $f_P = 1_P$.

Exercise 3.23 By using Theorem 3.2(2), show that for every abelian variety A, the morphism $A \to \text{Alb}(A)$ defined by the origin of A is an isomorphism of abelian varieties.

Here is a third formulation of the universal property of $\text{Alb}(X)$, which can be very useful:

Proposition 3.24 *Let X be a smooth projective connected k-variety with function field K, and let A be an abelian variety. Suppose that X has a rational* *point x (or more generally a zero-cycle of degree* 1). *Then we have a canonical isomorphism*

$$A(K)/A(k) \xrightarrow{\sim} \mathrm{Hom}_k(\mathrm{Alb}(X), A). \qquad (3.1.1)$$

Proof By Exercise 3.23, the functor Alb provides a canonical map

$$\mathrm{Mor}_k(X, A) \to \mathrm{Hom}_k(\mathrm{Alb}(X), A). \qquad (3.1.2)$$

The algebraic group structure of A gives the left-hand side an abelian group structure; since the addition $A \times A \to A$ is a homomorphism of abelian varieties, (3.1.2) is a homomorphism, which by functoriality factors as a homomorphism

$$\mathrm{Mor}_k(X, A)/\mathrm{Mor}_k(\mathrm{Spec}\, k, A) \to \mathrm{Hom}_k(\mathrm{Alb}(X), A). \qquad (3.1.3)$$

On the other hand, the Albanese morphism $X \to \mathrm{Alb}(X)$ sending x to 0 provides a homomorphism in the opposite direction

$$\mathrm{Hom}_k(\mathrm{Alb}(X), A) \to \mathrm{Mor}_k(X, A)$$

and one checks that the composition with the projection

$$\mathrm{Mor}_k(X, A) \to \mathrm{Mor}_k(X, A)/\mathrm{Mor}_k(\mathrm{Spec}\, k, A)$$

is inverse to (3.1.3). The proof is completed by using Theorem 3.17 to convert $\mathrm{Mor}_k(X, A)$ to $A(K)$. $\qquad \square$

Remark 3.25 Proposition 3.24 is false in general if X doesn't have any zero-cycle of degree 1 (example: X a nontrivial torsor with respect to an elliptic curve E without complex multiplication, $A = E$).

Theorem 3.26 (Picard–Albanese duality)

(1) *If P is a torsor with respect to the abelian variety A, then $\mathrm{Pic}^0_{P/k}$ is canonically isomorphic to $\mathrm{Pic}^0_{A/k} = A^*$.*
(2) *Let X and $(\mathrm{Alb}(X), P_X, f_X)$ be as in Theorem 3.21. Then the homomorphism*

$$f_X^* : \mathrm{Pic}(P_X) \to \mathrm{Pic}(X)$$

induces an isomorphism

$$\mathrm{Alb}(X)^* \simeq \mathrm{Pic}^0_{P_X/k} \xrightarrow{\sim} \mathrm{Pic}^0_{X/k}.$$

Theorem 3.27 *Every abelian variety A is the quotient of the Jacobian of a curve.*

Sketch (cf. [84, §10]). Choose a projective embedding $A \hookrightarrow \mathbf{P}^N$ given by a very ample line bundle on A. If k is infinite, Bertini's theorem provides us with an iterated hyperplane section C of A, passing through 0, which is a smooth projective connected curve. (If k is finite, we can also get such a curve C defined over k, replacing $\mathbf{P}^N$ by a Veronese embedding if necessary, i.e. cutting A by a complete intersection of sufficiently large multidegree.) We thus obtain a homomorphism

$$J(C) = \text{Alb}(C) \to A$$

and a connectivity lemma [84, lemma 10.3] implies surjectivity. $\qquad\square$

Remark 3.28 When the theory of abelian varieties was developed in the 1950s, people first constructed the Albanese variety, then the Picard variety as dual of the Picard variety as in Theorem 3.26: see for example Lang [Lan1], in particular ch. VI, th. 1. Then came Grothendieck, who revolutionised this by directly proving the representability of the Picard functor: see [BLR] for a detailed exposition of his method.

3.1.6 Polarisations; the Rosati involution

Let A be an abelian variety and let $\mathcal{L}$ be a line bundle on A. We have an associated map

$$\varphi_{\mathcal{L}} : A(k) \to \text{Pic}(A)$$
$$x \mapsto t_x^* \mathcal{L} \otimes \mathcal{L}^{-1}.$$

The theorem of the square ([Mum, § 6, cor. 4] or [83, th. 6.7]) implies that $\varphi_{\mathcal{L}}$ is a homomorphism whose image lies in $\text{Pic}^0(A)$. One shows that $\varphi_{\mathcal{L}}$ defines a homomorphism of abelian varieties $A \to A^*$, which is an isogeny if and only if $\mathcal{L}$ is ample.[3]

In particular, A and A^* are always isogenous.

Definition 3.29 A *polarisation* of A is an isogeny $\lambda : A \to A^*$ such that $\lambda_{\bar{k}}$ is of the form $\varphi_{\mathcal{L}}$ for a line bundle $\mathcal{L}$ over $A_{\bar{k}}$ (this line bundle may not be defined over k). We say that a polarisation is *principal* if it is an isomorphism.

Example 3.30 Suppose that $A = J(C)$, where C is a smooth projective curve of genus $g > 0$ with a rational point c; let $\iota : C \to A$ be the associated

---

[3] This is a somewhat distorted version of the theory: actually, one uses $\varphi_{\mathcal{L}}$ for $\mathcal{L}$ ample to construct the dual abelian variety A^* as a quotient of A by a finite group subscheme, cf. [Mum, p. 124] or [83, §§ 9, 10].

embedding (considering A as the Albanese variety of C). We let

$$\Theta_C = \iota(C) + \cdots + \iota(C) \quad (g-1 \text{ terms})$$

where the sum is relative to addition in A: this is the *associated theta divisor*. Then Θ_C defines a principal polarisation on A, independent of the choice of c [84, § 6].

Definition 3.31 Let λ be a polarisation of A. The *Rosati involution* associated to λ is the map from $\operatorname{End}^0(A)$ to itself given by

$$f \mapsto {}^\rho f = \lambda^{-1} f^* \lambda.$$

It is clear that the Rosati involution is an anti-automorphism of $\operatorname{End}^0(A)$; furthermore, we have

Proposition 3.32 *Let* $\lambda : A \to A^*$ *be a polarisation. Denote the pairing derived from the Weil pairing (definition 3.16) via the homomorphism* $T_l(\lambda) :$ *$T_l(A) \to T_l(A^*)$ by*

$$E^\lambda : T_l(A) \times T_l(A) \to \mathbf{Z}_l(1).$$

Then:

(i) E^λ *is antisymmetric.*
(ii) f *and* ${}^\rho f$ *are adjoint for* $E^\lambda \otimes \mathbf{Q}_l$, *for all* $f \in \operatorname{End}^0(A)$.
(iii) *The Rosati involution is an involution.*

Proof See [Mum, § 20, th. 1 and ff.] or [83, the beginning of § 17]. $\qquad\square$

Let R be a (finite-dimensional) semisimple algebra over a field K. We define a trace $\operatorname{Tr} : R \to K$ as follows:

(1) If R is simple with centre Z, then $\operatorname{Tr}(x) = \operatorname{Tr}_{Z/K} \operatorname{Trd}(x)$, where Trd is the reduced trace of R (over Z).
(2) If $R = R_1 \times \cdots \times R_n$, where the R_i are simple, then

$$\operatorname{Tr}(x_1, \ldots, x_n) = \operatorname{Tr}(x_1) + \cdots + \operatorname{Tr}(x_n).$$

The main result is the following.

Theorem 3.33 (Positivity of the Rosati involution) *Let* $\lambda : A \to A^*$ *be a polarisation. Then the quadratic form*

$$\operatorname{End}^0(A) \ni f \mapsto \operatorname{Tr}({}^\rho f f)$$

is positive definite. If $\lambda_{\bar{k}} = \varphi_{\mathcal{L}}$ and if $f \in \mathrm{End}(A)$, then

$$\mathrm{Tr}(^{\rho} f f) = \frac{2g}{(\mathcal{L})^g}((\mathcal{L})^{g-1} \cdot f^*\mathcal{L}).$$

Proof See [Mum, § 21] or [83, th. 17.3].   □

Corollary 3.34 *Let $\lambda : A \to A^*$ be a polarisation, let $f \in \mathrm{End}(A)$ satisfy $^{\rho} f f = a \in \mathbf{Z}$, and let $\alpha_1, \ldots, \alpha_{2g}$ be roots (in $\mathbf{C}$) of the characteristic polynomial of f (cf. Theorem 3.11). Then the subalgebra of $\mathrm{End}^0(A)$ generated by f is semisimple, and*

(i) $|\alpha_i|^2 = a$ *for all i;*
(ii) *the map $\alpha_i \mapsto a/\alpha_i$ is a permutation of the α_i.*

Proof This is an almost immediate consequence of the positivity theorem, see [Mum, § 21, Application II] or [83, lemma 19.3].   □

3.2 The Riemann hypothesis for an abelian variety

3.2.1 The zeta function of an abelian variety

Suppose now that $k = \mathbf{F}_q$ is finite. Let A be an abelian k-variety and let π_A be its Frobenius endomorphism. Then π_A acts as 0 on the tangent space at 0, so $1 - \pi_A^n$ is étale for all $n > 0$ and

$$A(\mathbf{F}_{q^n}) = \mathrm{Ker}(1 - \pi_A^n),$$

so

$$|A(\mathbf{F}_{q^n})| = P^{(n)}(1) = \prod_{i=1}^{2g}(1 - \alpha_i^n), \qquad (3.2.1)$$

where $P^{(n)}$ is the characteristic polynomial of π_A^n, with roots $\alpha_1^n, \ldots, \alpha_{2g}^n$.

Lemma 3.35 *For any Rosati involution, we have $^{\rho}\pi_A \pi_A = q$.*

Proof If λ is a polarisation, one can show that (since k is finite) $\lambda = \varphi_{\mathcal{L}}$ for a line bundle $\mathcal{L}$ defined over k. Then we have $\pi_A^* \mathcal{L} \simeq \mathcal{L}^q$ and the lemma follows from a small calculation (cf. [83, lem. 19.2]).   □

(It can be amusing to compare the proofs given in [Lan1], [Mum] and [83], which get simpler and simpler ...)

Theorem 3.36 (Weil)

(1) *The α_i satisfy $|\alpha_i| = \sqrt{q}$ for all i, and $\alpha_i \mapsto q/\alpha_i$ is a permutation of the roots of $P^{(1)}$.*

(2) *For $n \in \{0, \ldots, 2g\}$, let P_n be the polynomial whose inverse roots are the products $\alpha_{i_1} \ldots \alpha_{i_n}$ for $i_1 < \cdots < i_n$ (so $P_1 = P^{(1)}$). Then we have the factorisation*

$$Z(A, t) = \frac{P_1(t) \ldots P_{2g-1}(t)}{P_0(t) \ldots P_{2g}(t)}$$

and the functional equation

$$Z(A, 1/q^g t) = Z(A, t).$$

Proof This follows immediately from (3.2.1) and the formula from Proposition 2.3 (3). □

Remarks 3.37

(1) Before Tate modules were invented, Hasse, Deuring and Weil reasoned in terms of "l-adic matrices" derived from the action of the endomorphisms of A on torsion points, using the fact that

$$\mathrm{End}_{\mathbf{Z}}(\mathbf{Q}_l/\mathbf{Z}_l) = \mathbf{Z}_l.$$

(2) Once we know how to interpret the factorisation in Theorem 3.36 (2) in terms of a Weil cohomology H (3.6.3), we will get the following: the α_i are the eigenvalues of the Frobenius action on $H^1(A)$, and $H^*(A)$ is an exterior algebra on $H^1(A)$ (which can be deduced for all Weil cohomologies from the motivic structure of A [74]).

3.2.2 Connection with the case of curves

Let C, C' be smooth projective k-curves. Correspondences act on the Picard groups via the operation

$$\mathrm{Corr}(C, C') \times \mathrm{Pic}(C) \to \mathrm{Pic}(C')$$

$$(\gamma, D) \mapsto (p_{C'}^{C \times C'})_*(\gamma \cap D \times C')$$

(if the intersection product is defined; in the exceptional case $\gamma = P \times C'$, the second term is defined to be 0). This is in fact how Weil introduced the composition of correspondences.

This action maps $\mathrm{Pic}^0(C)$ to $\mathrm{Pic}^0(C')$; in fact, we have

Theorem 3.38 (Weil [Wei3, ch. VI, th. 22]) *If k is algebraically closed, this action induces an isomorphism*

$$\mathrm{Corr}_{\equiv}(C, C') \xrightarrow{\sim} \mathrm{Hom}_k(J(C), J(C')).$$

Proof (See [84, cor. 6.3] for a different proof.) By using the isomorphisms of Proposition 2.31 and Proposition 3.24, the representability of Pic^0 (Theorem 3.12) and the self-duality of the Jacobian of a curve (Example 3.30), we obtain the successive isomorphisms

$$\mathrm{Corr}_{\equiv}(C, C') \simeq \mathrm{Pic}^0(C'_{k(C)})/\mathrm{Pic}^0(C') \simeq J'(k(C))/J'(k) \xrightarrow{\sim} \mathrm{Hom}_k(J, J')$$

where $J = J(C)$ and $J' = J(C')$. One checks that the composed isomorphism is compatible with the action described above. $\square$

When $C' = C$, write $J = J(C)$ and let $\lambda = J \xrightarrow{\sim} J^*$ be the canonical principal polarisation of Example 3.30. One can show that the isomorphism

$$\mathrm{Corr}_{\equiv}(C, C) \xrightarrow{\sim} \mathrm{End}(J)$$

transforms the canonical involution of the first term (transposition) into the Rosati involution associated to λ, and that the trace σ from Definition 2.27 corresponds to the trace Tr of the semisimple algebra $\mathrm{End}^0(J)$ [Lan1, ch. VI, § 3, th. 6]. Thus, we recover Theorem 2.37 from Theorem 3.33[4]. We also see that the numerator of $Z(C, t)$ is the inverse characteristic polynomial of π_J.

Exercise 3.39 Generalise Theorem 3.38 to a split exact sequence

$$0 \to \mathrm{Pic}(X) \oplus \mathrm{Pic}(Y) \to \mathrm{Pic}(X \times Y) \to \mathrm{Hom}_k(\mathrm{Alb}(X), \mathrm{Pic}^0_{Y/k}) \to 0$$

for all smooth projective varieties X, Y over an algebraically closed field k. (Proceed as in the proof of Theorem 3.38; use the fact that the Néron–Severi group doesn't change upon extension of k.)

3.3 The Weil conjectures

The Weil conjectures are first stated at the end of [140]:

This, and other examples which we cannot discuss here, seem to lend some support to the following conjectural statements, which are known to be true for curves, but which I have not so far been able to prove for varieties of higher dimension.

---

[4] Weil himself deduced Theorem 3.33 from Theorem 2.37 [Wei3, n° 71]!

Let V be a variety[5] without singular points, of dimension n, defined over a finite field k with q elements. Let N_v be the number of rational points on V over the extension k_v of k of degree v. Then we have

$$\sum_1^\infty N_v U^{v-1} = \frac{d}{dU} \log Z(U)$$

where Z(U) is a rational function in U, satisfying a functional equation

$$Z(\frac{1}{q^n U}) = \pm q^{n\chi/2} U^\chi Z(U)$$

with χ equal to the Euler–Poincaré characteristic of V (the intersection number of the diagonal with itself on the product V × V).

Furthermore, we have:

$$Z(U) = \frac{P_1(U)P_3(U)\dots P_{2n-1}(U)}{P_0(U)P_2(U)\dots P_{2n}(U)}$$

with $P_0(U) = 1 - U$, $P_{2n}(U) = 1 - q^n U$, and, for $1 \le h \le 2n - 1$,

$$P_h(U) = \prod_{i=1}^{B_h} (1 - \alpha_{hi}U),$$

where the α_{hi} are algebraic integers of absolute value $q^{h/2}$.

Finally, let us call the degrees B_h of the polynomials $P_h(U)$ the Betti numbers *of the variety V; the Euler–Poincaré characteristic χ is then expressed by the usual formula $\chi = \sum_h (-1)^h B_h$. The evidence at hand seems to suggest that, if $\overline{V}$ is a variety without singular points, defined over a field K of algebraic numbers, then the Betti numbers of the varieties $V_\mathfrak{p}$, derived from $\overline{V}$ by reduction modulo a prime ideal $\mathfrak{p}$ in K, are equal to the Betti numbers of $\overline{V}$ (considered as a variety over the complex numbers) in the sense of combinatorial topology, for all except at most a finite number of prime ideals $\mathfrak{p}$. For instance, consider the Grassmannian variety...*

The case of Grassmannian varieties now seems easy, insofar as they are cellular. Curiously, Weil does not mention abelian varieties – the example he talks about at the beginning of his article is that of the generalised Fermat surfaces

$$H : \quad a_0 x_0^n + \cdots + a_r x_r^n = 0 \qquad (3.3.1)$$

for which he proved the conjecture by using Jacobi sums. He learned the value of the Betti numbers of H viewed as a complex manifold from Pierre Dolbeaut, which encouraged him to formulate his conjecture.

Exercise 3.40 (cf. [76, pp. 826–827]) Show that the Weil conjectures imply that the smallest constant γ appearing in Corollary 2.50 is equal to the degree

---

[5] Curiously, the word "projective" is absent.

of $P_{2n-1}(U)$. (This degree is equal to $2g$, where g is the dimension of the Picard variety of V, cf. remarks 3.37 (2) and 3.43.)

In comments from his scientific works about the paper, Weil writes:

I believe that I did some heuristic reasoning based on the Lefschetz formula. In classical algebraic geometry, let φ be a generically surjective map of degree d from a smooth, complete variety V of dimension n to itself. For each v, let N_v be the number of solutions, which we suppose to be finite, of $P = \varphi^v(P)$, or more precisely, the intersection number of the graph of the nth iteration of φ with the diagonal in $V \times V$. Arguments from classical combinatorial topology allow us to see that the function $Z(U)$ defined by (I)[6] satisfies a functional equation

$$Z(d^{-1}U^{-1}) = \pm(\sqrt{d}U)^\chi Z(U)$$

where χ is the Euler–Poincaré characteristic of V, or said otherwise (as I recalled in [140, p. 507]), χ is the intersection number of the diagonal with itself on $V \times V$. To hope from this that the same formula could apply over the finite base field $\mathbf{F}_q$ to the Frobenius map (for which we have $d = q^n$), and that my calculations would give the value of χ for the hypersurface (3.3.1), there was only one step remaining. When Dolbeault's result allowed me to verify this point, the inspection of these formulas suggested right away the conjectures on the Betti numbers and the "Riemann hypothesis"; the comparison with the known results over curves, as well as the tests of the Grassmanians and several other simple examples, didn't take long to confirm them.[7]

Weil didn't know it at the time, but Fermat hypersurfaces are "of abelian type": their motives can be expressed in terms of motives of abelian varieties (Katsura–Shioda [68]).

---

[6] this is the formula from Proposition 2.3 (3).

[7] *Je crois que je fis un raisonnement heuristique basé sur la formule de Lefschetz. En géométrie algébrique classique, soit φ une application génériquement surjective de degré d d'une variété V de dimension n, lisse et complète, dans elle-même. Pour chaque v, soit N_v le nombre de solutions, supposé fini, de $P = \varphi^v(P)$, ou pour mieux dire, le nombre d'intersection avec la diagonale, sur $V \times V$, du graphe de la n-ième itérée de φ. Des raisonnements de topologie combinatoire classique font voir alors que la fonction $Z(U)$ définie par (I) satisfait à une équation fonctionnelle*

$$Z(d^{-1}U^{-1}) = \pm(\sqrt{d}U)^\chi Z(U)$$

où χ est la caractéristique d'Euler-Poincaré de V, ou autrement dit (comme je le rappelais dans [140, p. 507]) le nombre d'intersection de la diagonale avec elle-même sur $V \times V$. De là à espérer que la même formule pouvait s'appliquer, sur le corps de base fini $\mathbf{F}_q$, à l'application de Frobenius (pour laquelle on a $d = q^n$), et que mes calculs donnaient donc la valeur de χ pour l'hypersurface (3.3.1), il n'y avait qu'un pas à faire. Quand le résultat de Dolbeault m'eut permis de vérifier ce point, l'inspection de mes formules suggéra de lui-même les conjectures sur les nombres de Betti et l'«hypothèse de Riemann»; la comparaison avec les résultats connus sur les courbes, ainsi que l'examen des Grassmanniennes et de quelques autres exemples simples, ne tarda pas à les confirmer.

3.4 Weil cohomologies

We begin with two quotes from Serre:

... the idea of counting points over $\mathbf{F}_q$ *by a Lefschetz formula was entirely* the idea of Weil. *I remember how enthusiastic I was when he explained it to me, and a few years later I managed to convey my enthusiasm to Grothendieck (whose taste was not a priori directed towards finite fields).* (Letter to S. Kleiman, 25 Mar. 1992, cited in [73, p. 8, note 3].)

... Weil formulated what were immediately called the Weil conjectures. *These conjectures apply to (projective, nonsingular) varieties over a finite field. They amount to supposing that the topological methods of Riemann, Lefschetz, Hodge... apply in characteristic* $p > 0$; *in this light, the number of solutions of an equation* (mod p) *appears as a number of fixed points, and must therefore be calculable by the Lefschetz trace formula. This truly revolutionary idea enthused mathematicians of the period (I can testify to this); it was the origin of a good part of the progress in algebraic geometry which followed.* [116, n° 5].[8]

Stepping back, we could consider Weil's idea to be the most direct ancestor of the philosophy of motives![9]

Serre and Grothendieck add some substance to this, as follows:

Definition 3.41 Let k be a field; consider the category $\mathbf{V}(k)$ of smooth projective varieties (morphisms: the k-morphisms). A *Weil cohomology* over $\mathbf{V}(k)$ is the data of a field K of characteristic zero and a (contravariant) functor

$$H^* : \mathbf{V}(k)^{\mathrm{op}} \to \mathbf{Vec}_K^*$$

from $\mathbf{V}(k)$ to the category of graded K-vector spaces, satisfying the following axioms:

(i) For all $X \in \mathbf{V}(k)$ of dimension n, the $H^i(X)$ are finite dimensional and vanish for $i \notin [0, 2n]$.

(ii) $H^0(\mathrm{Spec}\, k) = K$.

---

[8] *... Weil formule ce qu'on a tout de suite appelé les* conjectures de Weil. *Ces conjectures portent sur les variétés (projectives, non singulières) sur un corps fini. Elles reviennent à supposer que les méthodes topologiques de Riemann, Lefschetz, Hodge... s'appliquent en caractéristique* $p > 0$; *dans cette optique, le nombre de solutions d'une équation* (mod p) *apparaît comme un nombre de points fixes, et doit donc pouvoir être calculé par la formule des traces de Lefschetz. Cette idée, vraiment révolutionnaire, a enthousiasmé les mathématiciens de l'époque (je peux en témoigner de première main); elle a été à l'origine d'une bonne partie des progrès de la géométrie algébrique qui ont suivi.* [116, n° 5].

[9] Note that in [140] and the comments on this article, Weil doesn't explicitly mention the Lefschetz trace formula, or even singular (co)homology: he simply talks about "reasonings from classical combinatorial topology". Combinatorial topology is the old name of algebraic topology, the latter terminology was adopted by Bourbaki around 1944.

(iii) dim $H^2(\mathbf{P}^1) = 1$; we denote this vector space by $K(-1)$.

(iv) *Additivity*: if $X, Y \in \mathbf{V}(k)$, then the canonical morphism

$$H^*(X \coprod Y) \to H^*(X) \oplus H^*(Y)$$

is an isomorphism.

(v) *Normalisation*: if X is connected with field of constants E, then the canonical morphism

$$H^0(E) \to H^0(X)$$

is an isomorphism.

(vi) *Künneth formula*: for $X, Y \in \mathbf{V}(k)$, we have an isomorphism

$$\kappa_{X,Y} : H^*(X) \otimes H^*(Y) \xrightarrow{\sim} H^*(X \times Y)$$

which is natural in (X, Y), and verifies the obvious conditions on associativity, unity, and graded commutativity.

(Recall the last condition: if $x \in H^i(X)$ and $y \in H^j(Y)$, we ask that

$$x \otimes y = (-1)^{ij} y \otimes x$$

relative to the isomorphism $H^i(X) \otimes H^j(Y) \simeq H^j(Y) \otimes H^j(X)$ induced by the isomorphism $X \times Y \simeq Y \times X$.)

(vii) *Trace and Poincaré duality*: For all $X \in \mathbf{V}(k)$ of pure dimension n, we have a canonical morphism

$$\mathrm{Tr}_X : H^{2n}(X) \to K(-n) := K(-1)^{\otimes n}$$

which is an isomorphism if X is geometrically connected, and such that $\mathrm{Tr}_{X \times Y} = \mathrm{Tr}_X \otimes \mathrm{Tr}_Y$ modulo the Künneth formula; the "Poincaré pairing"

$$H^i(X) \otimes H^{2n-i}(X) \to H^{2n}(X \times X) \xrightarrow{\Delta_X^*} H^{2n}(X) \xrightarrow{\mathrm{Tr}_X} K(-n) \quad (3.4.1)$$

is nondegenerate.

(viii) *Cycle classes*: For all $X \in \mathbf{V}(k)$ and all $i \geq 0$, we have a homomorphism

$$\mathrm{cl}_X^i : CH^i(X) \to H^{2i}(X)(i) := \mathrm{Hom}_K(K(-i), H^{2i}(X))$$

where $CH^i(X)$ is the group of cycles of codimension i on X, modulo rational equivalence (the ith Chow group of X). These homomorphisms are contravariant in X and compatible with the Künneth formula ($\kappa_{X,Y}\left(\mathrm{cl}_X^i(\alpha) \otimes \mathrm{cl}_Y^j(\beta)\right) = \mathrm{cl}_{X \times Y}^{i+j}(\alpha \times \beta)$); furthermore, if $\dim X = n$, then the diagram

$$CH^n(X) \xrightarrow{\mathrm{cl}_X^n} H^{2n}(X)(n)$$

$$\deg \downarrow \qquad\qquad \mathrm{Tr}_X \downarrow$$

$$\mathbf{Z} \longrightarrow K$$

commutes.

Using the Künneth formula and the diagonal, we define the *cup product* on cohomology of $X \in \mathbf{V}(k)$:

$$\cdot : H^i(X) \times H^j(X) \to H^{i+j}(X) \qquad\qquad (3.4.2)$$

$$x \cdot y = \delta_X^*(\kappa_{X,X}(x \times y))$$

where $\delta_X : X \to X \times_k X$ is the diagonal morphism.

Lemma 3.42 *If E is a finite separable extension of k of degree d, then* $\dim H^0(E) = d$.

Proof Let $\tilde{E}$ be a Galois closure of E. By applying (iv) and (vi) to $E \otimes_k \tilde{E} \xrightarrow{\sim} \prod_{E \hookrightarrow_k \tilde{E}} \tilde{E}$, we get

$$H^0(E) \otimes H^0(\tilde{E}) \simeq H^0(\tilde{E})^d. \qquad \square$$

Remark 3.43 These axioms vary depending on the author. I have tried to take a minimalist approach, which should suffice for what follows. In particular:

- I have inserted axioms (iv) and (v), which I don't know how to deduce from the others.[10]
- I have omitted the "weak Lefschetz" and "strong Lefschetz" axioms from [71], which aren't necessary for proving the Weil conjectures.

An important additional axiom is the following (cf. Theorem 6.31): if X is projective, smooth, and geometrically connected, with a rational point x_0, then the Albanese morphism $X \to \mathrm{Alb}(X)$ associated to x_0 induces an isomorphism

$$H^1(\mathrm{Alb}(X)) \xrightarrow{\sim} H^1(X).$$

This axiom is satisfied by all the classical Weil cohomologies, i.e. those that appear in § 3.5.9.

Exercise 3.44 Show that

$$\mathrm{Tr}_{X \coprod Y} = \mathrm{Tr}_X + \mathrm{Tr}_Y$$

modulo the isomorphism (iv).

---

[10] Axiom (v) follows from (vii) if H factors through the functor $\mathbf{V}(k) \to \mathbf{V}(\bar{k})$, where $\bar{k}$ is an algebraic closure of k; but de Rham cohomology, which satisfies (v), doesn't have this property.

3.5 Formal properties of a Weil cohomology

Reference: Kleiman [71].

3.5.1 Direct image

Let $f : X \to Y$ be a morphism in $\mathbf{V}(k)$ with X, Y irreducible; let $\dim X = m$ and $\dim Y = n$. We define

$$f_* : H^i(X) \to H^{i+2n-2m}(Y)(n - m)$$

as the dual of

$$f^* : H^{2m-i}(Y)(m) \to H^{2m-i}(X)(m)$$

via Poincaré duality. We have $\mathrm{Tr}_X = (p_X)_*$ where p_X is the structural morphism of X, and the *projection formula*:

$$f_*(x \cdot f^* y) = f_* x \cdot y.$$

Let us prove it: by definition, f_* is adjoint to f^* for the Poincaré pairing. If z is an element of $H^*(Y)(*)$ of complementary dimension and weight, we have:

$$\langle f_*(x \cdot f^* y), z \rangle_Y = \langle x \cdot f^* y, f^* z \rangle_X = \langle x, f^*(y \cdot z) \rangle_X$$
$$= \langle f_* x, y \cdot z \rangle_Y = \langle f_* x \cdot y, z \rangle_Y.$$

3.5.2 Cohomological correspondences

Proposition 3.45 *Let $X, Y \in \mathbf{V}(k)$ be of dimension m and n respectively, and let $r \in \mathbf{Z}$. We have a canonical isomorphism*

$$\mathrm{Hom}^r(H^*(X), H^*(Y)) \simeq H^{2m+r}(X \times Y)(m)$$

where Hom^r denotes the "homogeneous homomorphisms of degree r".

Proof We have the sequence of isomorphisms

$$\mathrm{Hom}^r(H^*(X), H^*(Y)) = \prod_{i \geq 0} \mathrm{Hom}(H^i(X), H^{i+r}(Y))$$

$$\simeq \prod_{i \geq 0} H^i(X)^* \otimes H^{i+r}(Y) \overset{(*)}{\simeq} \prod_{i \geq 0} H^{2m-i}(X)(m) \otimes H^{i+r}(Y)$$

$$\simeq H^{2m+r}(X \times Y)(m)$$

where we have successively used the duality of K-vector spaces, Poincaré duality (axiom (vii)) and the Künneth formula (axiom (vi)).[11] $\qquad\square$

Definition 3.46 A *cohomological correspondence of degree r* from X^m to Y is an element of $H^{2m+r}(X \times Y)(m)$. The set of these is denoted $\mathrm{Corr}^r_H(X, Y)$. If X is not equidimensional, we extend this definition by additivity.

Proposition 3.45 defines *the composition of cohomological correspondences* by transport of structure. We can make this explicit (by the same strategy as used to prove the projection formula) as:

$$\beta \circ \alpha = (p_{13})_*(p_{12}^*\alpha \cdot p_{23}^*\beta) \tag{3.5.1}$$

for $\alpha \in \mathrm{Corr}^*_H(X_1, X_2)$ and $\beta \in \mathrm{Corr}^*_H(X_2, X_3)$, where p_{ij} denotes the projection from $X_1 \times X_2 \times X_3$ to the factor $X_i \times X_j$. These operate on cohomology by the formula

$$\alpha(x) = (p_Y)_*(p_X^*x \cdot \alpha) \tag{3.5.2}$$

for $\alpha \in \mathrm{Corr}^r_H(X, Y)$, $x \in H^i(X)$, $\alpha(x) \in H^{i+r}(Y)$.

Thus cohomological correspondences act on cohomology *covariantly*. For the sequel, let us spell out (3.5.2) in the case when α is given by a simple tensor $v \otimes w$ with $v \in H^{2m-i}(X)(m)$, $w \in H^{i+r}(Y)$ and where $x \in H^i(X)$:

$$\alpha(x) = \langle x, v \rangle w \tag{3.5.3}$$

(cf. [71, 1.3]).

3.5.3 Transposition

Poincaré duality gives an isomorphism

$$\mathrm{Hom}^r(H^*(X), H^*(Y)) \xrightarrow{\sim} \mathrm{Hom}^{2m-2n+r}(H^*(Y), H^*(X))(m-n)$$

($m = \dim X$, $n = \dim Y$); that is, an isomorphism

$$H^{2m+r}(X \times Y)(m) = \mathrm{Corr}^r_H(X, Y)$$
$$\xrightarrow{\sim} \mathrm{Corr}^{2m-2n+r}_H(Y, X)(m-n) = H^{2m+r}(Y \times X)(m)$$

called *transposition of correspondences*. We write $\varphi \mapsto {}^t\varphi$ for this isomorphism. For $v \otimes w$ in $H^{2m-i}(X)(m) \otimes H^{i+r}(Y)$ we have the Koszul sign rule:

$$(v \otimes w)' = (-1)^{(2m-i)(i+r)}w \otimes v = (-1)^{i(i+r)}w \otimes v. \tag{3.5.4}$$

---

[11] There is an ambiguity about the sign of the isomorphism (∗): we must choose the isomorphism $H^i(X)^* \xrightarrow{\sim} H^{2m-i}(X)(m)$ given by (3.4.1).

Example 3.47 If $f : Y \to X$ is a morphism, then $f^* : H^*(X) \to H^*(Y)$ defines a cohomological correspondence, denoted $f^* \in \mathrm{Corr}_H^0(X, Y)$. Its transpose is (by definition) f_*. We straightforwardly get the formulas $(g \circ f)^* = f^* \circ g^*$ and $(g \circ f)_* = g_* \circ f_*$. In particular,

$$\mathrm{Tr}_X = \mathrm{Tr}_Y \circ f_*. \qquad (3.5.5)$$

Lemma 3.48 *Let $\sigma_{X,Y} : Y \times X \xrightarrow{\sim} X \times Y$ exchange factors. Then for all $\varphi \in \mathrm{Corr}_H^r(X, Y)$, we have ${}^t\varphi = \sigma_{X,Y}^* \circ \varphi$.*

Proof It is in the same spirit as the proof of the projection formula (once again, ${}^t\varphi$ is defined to be the adjoint of φ under the Poincaré pairing). □

3.5.4 The Lefschetz trace formula

The Lefschetz trace formula is the following:

Proposition 3.49 *Let $\alpha \in \mathrm{Corr}_H^r(X, Y)$ and $\beta \in \mathrm{Corr}_H^{-r}(Y, X)$. We have:*

$$\langle \alpha, {}^t\beta \rangle_{X \times Y} = \sum_{i \geq 0} (-1)^i \mathrm{Tr}(\beta \circ \alpha \mid H^i(X))$$

where in the term on the right, $\mathrm{Tr}(\beta \circ \alpha \mid H^i(X))$ denotes the trace of $\beta \circ \alpha$ viewed as an endomorphism of the vector space $H^i(X)$.

In the case $\dim X = \dim Y = 1$, this formula goes back to Weil [Wei3, n° 43, 66 and 68].

Proof This is a trivial consequence of the Künneth formula and bilinearity; suppose $\alpha = v \otimes w$ and $\beta = w' \otimes v'$ with $v \in H^{2m-i}(X)(m)$, $w \in H^{i+r}(Y)$, $w' \in H^{2n-i-r}(Y)(n)$, $v' \in H^i(X)$. Then α (resp. β) vanishes outside of $H^i(X)$ (resp. of $H^{i+r}(Y)$) and, for $x \in H^i(X)$, $y \in H^{i+r}(Y)$, we have (3.5.3)

$$\alpha(x) = \langle x, v \rangle w, \qquad \beta(y) = \langle y, w' \rangle v',$$

so

$$\beta \circ \alpha(x) = \langle x, v \rangle \langle w, w' \rangle v'$$

and

$$\mathrm{Tr}(\beta \circ \alpha) = \langle v', v \rangle \langle w, w' \rangle.$$

Next, by using (3.5.4) we get:

$$\langle \alpha, {}^t\beta \rangle_{X \times Y} = (-1)^{i(i+r)} \langle v \otimes w, v' \otimes w' \rangle_{X \times Y}$$
$$= (-1)^{i(i+r)} \operatorname{Tr}_{X \times Y}(v \otimes w \cdot v' \otimes w')$$
$$= \operatorname{Tr}_{X \times Y}(v \cdot v' \otimes w \cdot w') \quad \text{(two signs cancel)}$$
$$= \langle v, v' \rangle \langle w, w' \rangle = (-1)^{i(2m-i)} \langle v', v \rangle \langle w, w' \rangle$$
$$= (-1)^i \operatorname{Tr}(\beta \circ \alpha). \qquad \square$$

Remark 3.50 Later we will see another, infinitely more conceptual, proof of the trace formula (Theorem 6.19 and Lemma A.37)!

Corollary 3.51 *If E/k is finite Galois with Galois group G, the representation*

$$G \to \operatorname{Aut}_K H^0(E)$$

given by the Galois action is the regular representation of G.

Proof Let $g \in G - \{1\}$: then the graph of the morphism $g : \operatorname{Spec} E \to \operatorname{Spec} E$ is transverse to that of the identity, so we have $\operatorname{Tr}(g) = \langle \Gamma_g, \Delta_{\operatorname{Spec} E} \rangle = 0$. This characterises the regular representation. $\qquad \square$

3.5.5 Chow groups: codimension and dimension

For more on Chow groups, see Fulton [Ful]. If $X \in \mathbf{V}(k)$ and $i \geq 0$, we write $CH^i(X)$ (resp. $CH_i(X)$) for the group of cycles of codimension (resp. of dimension) i on X, modulo rational equivalence. These have the following properties:

(1) The CH^* (resp. the CH_*) are contravariant (resp. covariant) functors on $\mathbf{V}(k)$.

(2) We have "intersection products"

$$CH^i(X) \times CH^j(X) \to CH^{i+j}(X).$$

(3) There is a degree homomorphism

$$\deg : CH_0(X) \to \mathbf{Z}.$$

(4) If X is purely of dimension n, we have

$$CH^i(X) = CH_{n-i}(X)$$

for all $i \in [0, n]$.

It is easy to prove covariant functoriality, as well as contravariant functoriality for flat morphisms; for arbitrary morphisms the latter is difficult to prove, as is the existence of intersection products.

3.5.6 Algebraic correspondences

In Section 2.9, we introduced algebraic correspondence between curves. This theory generalises to arbitrary smooth projective varieties.

Definition 3.52 A *Chow correspondence of degree r* from X^m to Y^n is an element of $CH^{m+r}(X \times Y)$. The set of these is denoted $\mathrm{Corr}^r(X, Y)$. If X is not equidimensional, we extend by additivity.

These correspondences compose according to the ever-recurring formula (3.5.1). They operate on Chow groups by the (similarly recurring) formula (3.5.2).

Remark 3.53 (3.5.2) is the particular case of (3.5.1) where $X_1 = \mathrm{Spec}\, k$.

Example 3.54 If $f : X \to Y$ is a morphism, its graph Γ_f defines a correspondence $f^* \in \mathrm{Corr}^0(Y, X)$.

Proposition 3.55

(1) *The composition of correspondences is associative, and the class of the diagonal is a left and right identity element.*
(2) *If $X \xrightarrow{f} Y \xrightarrow{g} Z$ are morphisms, we have $(g \circ f)^* = f^* \circ g^*$.*
(3) *The action of the correspondence f^* on Chow groups is the inverse image of cycles.*

Warning 3.56 For reasons of compatibility with the literature, we have changed the direction of composition of correspondences compared with 2.9! More precisely, there are two conventions for the composition of algebraic correspondences . The covariant convention is the one where the graph of a morphism operates in a covariant way: this is the convention used by Weil, by Fulton [Ful, ch. 16] and by Voevodsky [135, § 2.1]. Conversely, Grothendieck uses the contravariant convention to have a compatibility with the contravariance of cohomology. This is also the convention used by Kleiman [73] and Scholl [104], and the one that we take here. One can pass "trivially" from one convention to the other, but this carries a risk of confusion, cf. [66, 7.1] and the following.

The clearest way to understand this situation is to consider the action on the Chow groups: these correspondences are adapted to the CH^* theory, while the covariant correspondences are adapted to the CH_* theory. Thus we define the *covariant correspondences of degree r*

$$\mathrm{Corr}_r(X, Y) = CH_{m+r}(X \times Y)$$

(if X is purely of dimension m; we pass to the general case by additivity).

Their composition is defined once again by (3.5.1), and they operate on CH_* by (3.5.2), with Remark 3.53 still valid.

Here are two ways of passing from contravariant correspondences to covariant correspondences, and vice versa.

Lemma 3.57

(1) *The exchange of factors $\sigma_{X,Y}$ from Lemma* 3.48 *induces involutive isomorphisms called* transposition

$$\mathrm{Corr}^r(X, Y) \xrightarrow{\sim} \mathrm{Corr}_{-r}(Y, X) \xrightarrow{\sim} \mathrm{Corr}^r(X, Y)$$

denoted $\alpha \mapsto {}^t\alpha$. We have ${}^t(\beta \circ \alpha) = {}^t\alpha \circ {}^t\beta$.

(2) *If X (resp. Y) is purely of dimension m (resp. n), we have equality*

$$\mathrm{Corr}^r(X, Y) = \mathrm{Corr}_{n-m-r}(X, Y).$$

Example 3.58 If $f : X \to Y$ is a morphism, its graph Γ_f defines a covariant correspondence $f_* \in \mathrm{Corr}_0(X, Y)$, which operates on the Chow groups CH_* by direct image. We have ${}^t f^* = f_*$ and ${}^t f_* = f^*$.

Fulton proves the covariant version of Proposition 3.55 in [Ful, prop. 16.1.1]. We can also deduce this from Lemma 3.57 (1) (which is easy), or by simply mimicking the computations in loc. cit.

3.5.7 Algebraic and cohomological correspondences

The cycle class defines homomorphisms

$$\mathrm{cl} : \mathrm{Corr}^r(X, Y) \to \mathrm{Corr}_H^{2r}(X, Y)(r).$$

Note the change of degree and the twist!

Lemma 3.59

(1) cl *respects composition of correspondences.*

(2) *For all algebraic correspondences α, we have*

$$\mathrm{cl}(^t\alpha) = {}^t\,\mathrm{cl}(\alpha).$$

In particular, for all morphisms $f : X \to Y$, we have $\mathrm{cl}\circ f_* = f_* \circ \mathrm{cl}$ (direct image in cohomology).

Proof (1) follows directly from formula (3.5.1) and the axioms of a Weil cohomology, while (2) follows from Lemma 3.48. □

Corollary 3.60 *Suppose X and Y are equidimensional, of respective dimensions m and n. Then the diagram*

$$
\begin{array}{ccc}
CH^i(X) & \xrightarrow{\mathrm{cl}_X^i} & H^{2i}(X)(i) \\
f_* \downarrow & & f_* \downarrow \\
CH^{i+n-m}(Y) & \xrightarrow{\mathrm{cl}_Y^{i+n-m}} & H^{2i+2n-2m}(Y)(i+n-m)
\end{array}
$$

commutes for all $i \geq 0$.

3.5.8 First applications of the trace formula

By applying the trace formula to the identity correspondence, we first get:

Proposition 3.61 *For all $X \in \mathbf{V}(k)$, we have*

$$\chi(X) = \sum_{i\geq 0}(-1)^i \dim H^i(X)$$

where the ith term on the left is the Euler–Poincaré characteristic of X, defined as the self-intersection number of the diagonal (cf. § 3.3). □

Corollary 3.62 *Let H^* be a Weil cohomology. Then for all smooth projective geometrically connected k-curves C of genus g, we have*

$$\dim H^1(C) = 2g.$$

Proof Following (the proof of) Proposition 2.36, we have $\chi(C) = 2-2g$. By applying Proposition 3.61, we obtain that

$$2 - 2g = \dim H^0(C) - \dim H^1(C) + \dim H^2(C)$$

from which the corollary follows by applying axiom (v). □

3.5.9 Classical Weil cohomologies

The primordial example of a Weil cohomology is *Betti cohomology*, when $k = \mathbf{C}$:

$$H_B^i(X) = H^i(X(\mathbf{C}), \mathbf{Q})$$

where the term on the right is singular cohomology. At the time Weil made his conjectures, this was the only such cohomology known, in characteristic zero.

In characteristic p, we cannot expect to have any Weil cohomology with coefficients in $\mathbf{Q}$, or even in $\mathbf{R}$ (an observation of Serre). The obstruction to this is Corollary 3.62 applied to a *supersingular* elliptic curve E.

By definition, an elliptic curve over k is supersingular if $E(\bar{k})$ has no p-torsion, where $p = \operatorname{char} k$: Deuring [35, p. 198] observed that these curves always exist over finite fields, and that the ring $\operatorname{End}^0(E_{\bar{k}})$ is a quaternionic field with centre $\mathbf{Q}$, ramified precisely at p and ∞. Now $\operatorname{End}^0(E) \otimes K$ acts on $H^1(E)$ via action of correspondences. But a quaternion algebra with centre K cannot operate on a K-vector space of dimension 2 unless it is split…

Étale cohomology gave birth to *l-adic cohomology*

$$H_l^i(X) = H^i(X_{\bar{k}}, \mathbf{Q}_l) := \varprojlim H_{\text{ét}}^i(X_{\bar{k}}, \mathbf{Z}/l^n) \otimes_{\mathbf{Z}_l} \mathbf{Q}_l$$

for all primes $l \neq \operatorname{char} k$. Its coefficients are $\mathbf{Q}_l$. In characteristic zero, we also have *algebraic de Rham cohomology*

$$H_{\text{dR}}^i(X) = \mathbb{H}_{\text{Zar}}^i(X, \Omega_{X/k}^*),$$

the Zariski hypercohomology of the complex of Kähler differential forms, with coefficients in k. Over a perfect field k of characteristic p, we have *crystalline cohomology*

$$H_{\text{crys}}^i(X)$$

whose coefficients are the field of fractions of $W(k)$, the ring of Witt vectors of k. If $k = \mathbf{F}_q$ and E is a supersingular elliptic curve, then $\operatorname{End}^0(E) = \operatorname{End}^0(E_{\bar{k}})$ only if $q = p^n$ with n even; in this case, the field of fractions of $W(k)$ is an even degree extension of $\mathbf{Q}_p$, which splits $\operatorname{End}^0(E)$!

These cohomologies are sometimes called *classical Weil cohomologies*.

Exercise 3.63 For all classical Weil cohomologies H, compute the commutative algebra $H^0(E)$, where E is a finite extension of k.

Exercise 3.64 We fix a base field K.

(a) Let H be a Weil cohomology with coefficients in K. Then for any extension L of K, $X \mapsto H^*(X) \otimes_K L$ defines a Weil cohomology with coefficients in L.

(b) For all fields F of characteristic 0, there is a Weil cohomology with coefficients in an extension of F.

3.6 Proofs of some of the Weil conjectures

The most important application of the trace formula is the following theorem.

Theorem 3.65 *If there exists a Weil cohomology on* $\mathbf{V}(\mathbf{F}_q)$ *with coefficients in* K, *then the Weil conjectures concerning rationality, the functional equation, and factorisation* (*into polynomials with coefficients in* K) *all hold.*

Proof Let $X \in \mathbf{V}(k)$ be geometrically connected and of dimension n. We apply the trace formula to powers of the Frobenius endomorphism π_X considered as a cohomological correspondence:

$$\langle \Delta_X, (\pi_X^r)^* \rangle_{X \times X} = \sum_{i=0}^{2n} (-1)^i \operatorname{Tr}((\pi_X^r)^* \mid H^i(X)). \tag{3.6.1}$$

As in § 2.9.9, the left-hand side is equal to $|X(\mathbf{F}_{q^r})|$. Rationality and the factorisation of $Z(X, t)$ follow from the identity:

$$\exp \left(\sum_{r=0}^{\infty} \frac{\operatorname{Tr}(f^r)}{r} t^r \right) = \frac{1}{\det(1 - ft \mid V)} \tag{3.6.2}$$

where f is an endomorphism of a finite-dimensional K-vector space V,[12] whence

$$Z(X, t) = \prod_{i=0}^{2n} P_i(t)^{(-1)^{i+1}}, \quad P_i(t) = \det(1 - \pi_X^* t \mid H^i(X)). \tag{3.6.3}$$

We see then that Weil's "Betti numbers" are indeed given by

$$B_i = \dim H^i(X).$$

This gives that $Z(X, t)$ is a rational function *a priori* with coefficients in K; but a criterion for rationality of formal power series using Hankel

---

[12] To prove this identity, we may reduce to the case of algebraically closed K, then put f in triangular form, and finally reduce to the case $\dim V = 1$ and f is a scalar.

determinants [Ami, Prop. 5.2.1] (see also Exercise 3.71) allows us to see that $\mathbf{Q}[[t]] \cap K(t) = \mathbf{Q}(t)$. However, it doesn't allow us to show that the $P_i(t)$ are individually in $\mathbf{Q}[t]$, without knowing how to separate their roots.

To obtain the functional equation, we use Poincaré duality: the operator π_X^* is almost unitary with respect to the Poincaré pairing. More precisely, by comparing (3.6.3) for $X = \mathbf{P}^1$ with the formula from Proposition 2.3 (4), we see that $\pi_{\mathbf{P}^1}^*$ operates on the one-dimensional vector space $H^2(\mathbf{P}^1)$ by multiplication by q. By axiom (vii) of a Weil cohomology, π_X^* acts on $H^{2n}(X)$ by multiplication by q^n.[13] We then have, for $x \in H^i(X)$ and $y \in H^{2n-i}(X)$:

$$< \pi_X^* x, \pi_X^* y > = \mathrm{Tr}_X(\pi_X^*(x \cdot y)) = q^n \, \mathrm{Tr}_X(x \cdot y) = q^n < x, y > . \quad (3.6.4)$$

We deduce:

$$\begin{aligned}
P_{2n-i}(1/q^n t) &= \det(1 - \pi_X^*/q^n t \mid H^{2n-i}(X)) \\
&= \det(1 - (\pi_X^*)^{-1} t^{-1} \mid H^i(X)) \\
&= \det(\pi_X^* \mid H^i(X))^{-1} (-t)^{-B_i} P_i(t). \quad (3.6.5)
\end{aligned}$$

But by reapplying (3.6.4), we find

$$\det(\pi_X^* \mid H^i(X)) \det(\pi_X^* \mid H^{2n-i}(X)) = q^{nB_i}.$$

By collecting (3.6.5) for $i = 0, \ldots, 2n$, noting that $B_i = B_{2n-i}$ by Poincaré duality and considering Proposition 3.61, we obtain the desired functional equation:

$$Z(X, 1/q^n t) = \varepsilon q^{n\chi/2} (-t)^\chi Z(X, t) \quad (3.6.6)$$

with a sign ε that arises from the central cohomology group $H^n(X)$. $\qquad \square$

Remark 3.66 This sign may be -1! A simple example of this is the blow-up of $\mathbf{P}_{\mathbf{F}_q}^2$ at the quadratic closed point $\{[0 : 1 : \sqrt{d}], [0 : 1 : -\sqrt{d}]\}$ where d is not a square in $\mathbf{F}_q$, cf. [60, ex. 14.6 c)]. (As Serre pointed out to me, an even simpler example is that of a smooth quadric surface of discriminant d over $\mathbf{F}_q$.)

Exercise 3.67 Show that the integer $n\chi$ from the functional equation (3.6.6) is always even.

---

[13] To justify this, choose a dominant rational map $f : X \dashrightarrow (\mathbf{P}^1)^n$. The graph of f induces an isomorphism $f_* : H^{2n}(X) \xrightarrow{\sim} H^{2n}((\mathbf{P}^1)^n) \simeq H^2(\mathbf{P}^1)^{\otimes n}$, and π_X^*, $\pi_{(\mathbf{P}^1)^n}^*$ commute with f_* while the last isomorphism transforms $\pi_{(\mathbf{P}^1)^n}^*$ into $(\pi_{\mathbf{P}^1}^*)^{\otimes n}$. We will later give (Proposition 6.46) a more reasonable proof of this fact.

3.6.1 The other Weil conjectures

Two of these conjectures are not addressed by the formalism of Weil cohomology: the value of the Betti numbers when X is the reduction of a variety defined on a field of characteristic zero, and the Riemann hypothesis.

The first of these was proved using properties of l-adic cohomology, namely,

l-adic-Betti comparison theorem (M. Artin).

If $X \in \mathbf{V}(\mathbf{C})$, we have canonical and functorial isomorphisms in X

$$H_B^i(X) \otimes_{\mathbf{Q}} \mathbf{Q}_l \simeq H_l^i(X).$$

Invariance under a separably closed extension.

If K/k is a separably closed field extension of characteristic $\neq l$, then

$$H_l^i(X) \xrightarrow{\sim} H_l^i(X_K)$$

is an isomorphism for all $X \in \mathbf{V}(k)$.

Smooth and proper base change.

Let A be a discrete valuation ring with field of fractions K and residue field k. Let $\mathcal{X}$ be a smooth projective A-scheme with generic fibre X and special fibre X_0. Then we have a canonical and functorial isomorphism in $\mathcal{X}$:

$$H_l^i(X) \simeq H_l^i(X_0).$$

For the second conjecture, the natural strategy would be to prove a generalisation of Weil's "important lemma": an inequality of the type

$$\mathrm{Tr}(^t\pi_X \circ \pi_X) > 0.$$

This has meaning over $H^n(X)$ if $\dim X = n$, but what about over the other H^i? Serre gives an answer using a polarisation in the context of Kähler manifolds using the Hodge index theorem (in Hodge theory) [109]. These arguments irresistibly evoke Hilbert and Pólya's idea for proving the classical Riemann hypothesis...

Serre's identity gave birth to the *standard conjectures* of Grothendieck and Bombieri [44]; a programme for proving the Riemann hypothesis over finite fields. Unfortunately, this approach is still at an impasse: the standard conjectures have yet to be proved.

It was Deligne who finally proved the last of the Weil conjectures in 1974 [26]. He gives a second proof in [30]. Since then, at least two other proofs have been given: those of Laumon [78] and of Kedlaya [70].

Let us note the following elementary fact.

Lemma 3.68 *If X satisfies the last Weil conjecture, then the polynomials P_i are pairwise coprime, with integer coefficients independent of the choice of l.*

Proof The first statement is clear, since the roots of P_i and of P_j have different complex absolute values when $i \neq j$. As $Z(X,t) \in \mathbf{Q}(t)$, the group $\mathrm{Gal}(\bar{\mathbf{Q}}/\mathbf{Q})$ permutes the set of zeros and poles; once again, because of the Riemann hypothesis, it leaves the roots of P_i stable for all i, and so $P_i \in \mathbf{Q}[t]$. Such a factorisation of $Z(X,t)$ doesn't depend on the choice of l.

Finally, if the inverse roots of P_i are algebraic integers, then P_i has integer coefficients. $\qquad\square$

3.7 Dwork's theorem

Theorem 3.69 (Bernard Dwork, [38]) *Let X be a scheme of finite type over $\mathbf{F}_q$. Then $Z(X,t) \in \mathbf{Q}(t)$.*

A very pleasant description of Dwork's proof is given by Serre in his Bourbaki talk [110]. I reproduce the main points:

a) By standard dévissage arguments (cf. proof of Theorem 2.5), Dwork reduces to the case of a "toric" hypersurface (the intersection of a hypersurface of $\mathbf{A}^n$ and the open set $\prod t_i \neq 0$). We may further suppose that $q = p$.

b) He generalises an elegant theorem of Emile Borel [15] as follows.

Theorem 3.70 (Borel–Dwork Theorem) *Let $f(t) = \sum_{n=0}^{\infty} a_n t^n$ be a power series with integer coefficients. Suppose there exist two real numbers r_∞, $r_p > 0$ such that*

(i) *f is meromorphic in the disc $|t| < r_\infty$ of $\mathbf{C}$.*
(ii) *f is meromorphic in the disc $|t| < r_p$ of $\mathbf{C}_p$.*
(iii) *$r_\infty r_p > 1$.*

Then $f \in \mathbf{Q}(t)$.

Here, $\mathbf{C}_p$ is the completion of an algebraic closure of $\mathbf{Q}_p$ and "meromorphic" has essentially the same meaning as on $\mathbf{C}$. The proof is not very difficult: cf. [Ami, th. 5.3.2].

c) We apply Theorem 3.70 to $f(t) = Z(X,t)$. As this function is dominated by $Z(\mathbf{A}^n, t) = 1/(1 - p^n t)$, we may take $r_\infty = p^{-n}$ (which is even holomorphic in this domain), and one must check that we can take $r_p > p^n$. In fact, we may take $r_p = +\infty$: Dwork established a factorisation (cf. [110, (48)])

$$Z(X, pt) = \prod_{i=0}^{n} (1 - p^{n-i}t)^{(-1)^{i+1}\binom{n}{i}} \prod_{i=0}^{n+1} \Delta(p^{n+1-i}t)^{(-1)^{i+1}\binom{n+1}{i}} \quad (3.7.1)$$

where

$$\Delta(t) = \det(1 - t\Psi)$$

has an infinite radius of convergence in $\mathbf{C}_p$. Here, Ψ denotes an endomorphism of an infinite-dimensional $\mathbf{C}_p$-vector space (more precisely, the limit of a sequence of such operators), whose nature was later clarified by Serre in [111] (*p*-adic Fredholm theory). Formula (3.7.1) is then interpreted as arising from a Koszul resolution (loc. cit., p. 85).

Dwork's proof was reinterpreted a little later using a cohomology theory, namely the Monsky–Washnitzer cohomology, which then gave rise (much later) to "*p*-adic cohomology" or "rigid cohomology": see Arabia–Mebkhout [6] and Le Stum [LS].

Exercise 3.71 Let K be a field. Suppose we are given a formal power series $f = \sum_{n=0}^{\infty} a_n t^n$ with coefficients in K.

(a) Show that the following conditions are equivalent for all $k \geq 0$:
   (i) There exists a polynomial $P \in K[t]$ of degree k such that $Pf \in K[t]$.
   (ii) The sequence (a_n) satisfies a linear recurrence relation

   $$a_{n+k} + b_1 a_{n+k-1} + \cdots + b_k a_n = 0, \quad b_1, \ldots, b_k \in K$$

   for $n \geq n_0$, and suitable n_0.
(b) If k is minimal in (a), show that the b_i are unique.
(c) Let L be a field extension of K and suppose that $f \in L(t)$. Show that $f \in K(t)$. (Choose a basis of the K-vector space L containing 1. The reader who does not like the axiom of choice may suppose that L is of finite type over K, and reduce from there.)

(This method is less computational than using Hankel determinants [Ami, §5.2].)

4

L-functions from number theory

4.1 Dirichlet *L*-functions

4.1.1 Two approaches towards Euler products

The approaches are the following:

(i) One works up to a finite number of factors.
(ii) One insists on having precise results, thus all the factors come into play.

The first is often enough for a first approximation, but one would like to then pass to the second. We begin by illustrating this for Dirichlet characters.

4.1.2 Dirichlet characters and modular characters

Definition 4.1 a) A *Dirichlet character modulo m* is a function $\chi : \mathbf{Z} \to \mathbf{C}$ such that

(i) $\chi \neq 0$.
(ii) $\chi(x + m) = \chi(x)\ \forall x \in \mathbf{Z}$ (χ is periodic of period m).
(iii) $\chi(xy) = \chi(x)\chi(y)\ \forall x, y \in \mathbf{Z}$.
(iv) $\chi(x) = 0$ if $(x, m) \neq 1$.

b) A *modular character modulo m* is a character of the abelian group $(\mathbf{Z}/m)^*$ (a homomorphism from $(\mathbf{Z}/m)^*$ to $\mathbf{C}^*$).

Remark 4.2 A Dirichlet character modulo m is a character of $(\mathbf{Z}/m)^*$ extended by 0 to $\mathbf{Z}/m$ and lifted to $\mathbf{Z}$. If $m'|m$, then every character of $(\mathbf{Z}/m')^*$ defines a character of $(\mathbf{Z}/m)^*$ via $(\mathbf{Z}/m)^* \to (\mathbf{Z}/m')^*$. But the associated Dirichlet characters are different in general. For example, $\chi = 1 \pmod{m}$ corresponds to

73

$$1_m : \mathbf{Z} \longrightarrow \mathbf{C}$$
$$x \longmapsto \begin{cases} 1 & \text{if } (x, m) = 1; \\ 0 & \text{otherwise.} \end{cases}$$

We are then led to

Definition 4.3 Let χ be a Dirichlet character modulo m, seen as a function from $\mathbf{Z}/m$ to $\mathbf{C}$. The *conductor* f of χ is the smallest divisor of m such that $\chi|_{(\mathbf{Z}/m)^*}$ factors through $(\mathbf{Z}/f)^*$. We say that χ is *primitive* if its conductor is equal to m.

Exercise 4.4 Prove the existence of f.

4.1.3 Dirichlet *L*-functions

Definition 4.5 Let $m \in \mathbf{N} - \{0\}$ and let χ be a Dirichlet character modulo m. We define the *Dirichlet L-function* (associated to χ) to be the series

$$L(\chi, s) = \sum_{n=1}^{\infty} \frac{\chi(n)}{n^s}.$$

Since χ is completely multiplicative, Proposition 1.13 gives a factorisation as an Euler product:

Proposition 4.6 *For all χ, we have the identity (of formal Dirichlet series)*

$$L(\chi, s) = \prod_{p \in \mathcal{P}} \frac{1}{1 - \chi(p)p^{-s}},$$

where $\mathcal{P}$ denotes the set of prime numbers.

Let us compare the *L*-function associated to a Dirichlet character and the one associated to the corresponding primitive character.

Proposition 4.7 *Let χ be a Dirichlet character modulo m, f its conductor, and finally let χ' be the associated primitive character modulo f. Then*

$$L(\chi, s) = L(\chi', s) \prod_{p \in S} \left(1 - \frac{\chi'(p)}{p^s}\right),$$

where S denotes the set of prime factors of m not dividing f. In particular:

(1) *For* $\chi = 1_m$, *we have* $L(1_m, s) = F(s)\zeta(s)$, *where* $F(s) = \prod_{p|m} \left(1 - \frac{1}{p^s}\right)$.

(2) *If* m *and* f *have the same prime factors, then* $L(\chi, s) = L(\chi', s)$.

4.1.4 *L*-functions and zeta functions

Let $K = \mathbf{Q}(\mu_m)$: this is a Galois extension of $\mathbf{Q}$, with Galois group G canonically isomorphic to $(\mathbf{Z}/m)^*$. Recall this isomorphism: the cyclotomic character

$$\kappa : G \to (\mathbf{Z}/m)^*$$

defined by $g\zeta = \zeta^{\kappa(g)}$ for all $\zeta \in \mu_m$ is clearly injective, and a theorem of Gauss implies that it is surjective (by the irreducibility of cyclotomic polynomials: see a beautiful proof in Cassels–Fröhlich [CF, ch. III, lemma 1]). Its degree is therefore $\varphi(m)$, the Euler totient of m.

The Dirichlet L-functions allow us to factorise the zeta function of K:

Proposition 4.8 *We have*

$$\zeta_K(s) = G(s) \prod_{\chi} L(\chi, s)$$

where the product runs over all characters χ *of* $(\mathbf{Z}/m\mathbf{Z})^*$ *and*

$$G(s) = \prod_{\mathfrak{P}|m} \left(1 - \frac{1}{N(\mathfrak{P})^s}\right)^{-1}$$

(a finite product of Eulerian factors).

Proof If $p \nmid m$, we simply write p for its image in $(\mathbf{Z}/m\mathbf{Z})^*$. We also write $f(p)$ for the order of p in this group, and $g(p) = \frac{\varphi(m)}{f(p)}$ for the order of $(\mathbf{Z}/m\mathbf{Z})^*/\langle p \rangle$. Then $f(p)$ is the residual degree of p in the extension $K/\mathbf{Q}$, and $g(p)$ is the number of prime ideals of K over p. Hence,

$$\prod_{\mathfrak{P}|p} \left(1 - \frac{1}{N(\mathfrak{P})^s}\right)^{-1} = \frac{1}{(1 - \frac{1}{p^{f(p)s}})^{g(p)}}.$$

The statement then follows from the easy identity

$$\prod_{\chi}(1 - \chi(p)t) = (1 - t^{f(p)})^{g(p)},$$

where χ runs over the characters of $(\mathbf{Z}/m)^*$. $\qquad\square$

We see that by naïve considerations with the Dirichlet characters, we arrive at only an approximate expression for $\zeta_K(s)$, which avoids precisely the local factors corresponding to ramified prime ideals. To obtain an exact value we must replace each character with the associated primitive character, and we obtain

Theorem 4.9 *For any integer $n > 1$, write* $\mathrm{Prim}(n)$ *for the set of primitive modular characters modulo n. Then if $K = \mathbf{Q}(\mu_m)$, we have*

$$\zeta_K(s) = \prod_{d \mid m} \prod_{\chi \in \mathrm{Prim}(d)} L(\chi, s).$$

(The proof is essentially an extension of that of Proposition 4.8 to the case of ramified primes: recall that p is ramified in $K/\mathbf{Q}$ if and only if it divides m, and then its index of ramification is $\varphi(p^r)$ where p^r is the greatest power of p dividing m, cf. [CF, ch. III].)

See Exercise 4.64 (d) for a generalisation of Theorem 4.9 using Artin L-functions.

4.2 The Dirichlet theorems

4.2.1 Convergence of generalised Dirichlet series

This paragraph is taken from [Ser3, ch. VI, § 2.2]. Its goal is to collect general results on convergence of Dirichlet series, which go back essentially to Landau.

Theorem 4.10 *Let $(\lambda_n)_{n \geq 1}$ be a strictly increasing sequence of real numbers tending to $+\infty$, and let $(a_n)_{n \geq 1}$ be a sequence of complex numbers. If the series $f(s) = \sum a_n e^{-\lambda_n s}$ converges for $s = z_0 \in \mathbf{C}$, it converges uniformly in the domain $\Re(s - z_0) \geq 0$, $|\mathrm{Arg}(s - z_0)| \leq \alpha$ for all $\alpha \in (0, \pi/2)$. We have $f(s) \longrightarrow f(z_0)$ if $s \longrightarrow z_0$ in such a domain. In particular, f converges in the domain $\Re(z) > \Re(z_0)$, where it defines a holomorphic function.*

Right away we note the two most interesting special cases:

- $\lambda_n = n$: we recover power series up to the change of variables $z = e^{-s}$.
- $\lambda_n = \log n$: we recover the usual Dirichlet series.

Proof Up to a translation on s, we may suppose that $z_0 = 0$. The hypothesis thus shows that the series $\sum a_n$ converges. We first must show that we have uniform convergence in any domain of the form $\Re(s) \geq 0$, $\left| \frac{s}{\Re(s)} \right| \leq k$.

Let $\varepsilon > 0$. Since $\sum a_n$ converges, there exists an $N \in \mathbf{N}$ such that, for all $m, m' \geq N$, we have $|A_{m,m'}| \leq \varepsilon$ with $A_{m,m'} = \sum_{n=m}^{m'} a_n$. If we apply Lemma 1.17 with $b_n = e^{-\lambda_n s}$, we find:

$$S_{m,m'} = \sum_{n=m}^{m'-1} A_{m,n}(e^{-\lambda_n s} - e^{-\lambda_{n+1} s}) + A_{m,m'}e^{-\lambda_{m'} s}.$$

Set $s = x + iy$ and apply Lemma 1.16. But then

$$|S_{m,m'}| \leq \varepsilon \left(1 + \frac{|s|}{x} \sum_{n=m}^{m'-1} (e^{-\lambda_n x} - e^{-\lambda_{n+1} x})\right)$$

where $|S_{m,m'}| \leq \varepsilon(1 + k(e^{-\lambda_m x} - e^{-\lambda_{m'} x})) \leq \varepsilon(1 + k)$, giving uniform convergence. The fact that $f(z) \to f(z_0)$ then follows. To show that f is holomorphic, we apply Lemma 1.3 to the partial sums $f_n(s) = \sum_{i \leq n} a_i e^{-\lambda_i s}$. $\square$

Corollary 4.11 *With the notation from Theorem 4.10, the domain of convergence of f contains a maximal open half-plane.* $\square$

Corollary 4.12 *If f is identically zero, then $a_n = 0$ for all $n > 0$. In particular, a non-zero Dirichlet series gives rise to a holomorphic function which is not identically zero in its domain of convergence.*

Proof Suppose the contrary and let n be the smallest integer such that $a_n \neq 0$. We multiply f by $e^{\lambda_n s}$, and let $s \in \mathbf{R}$ tend to $+\infty$. Uniform convergence tells us that $e^{\lambda_n s} f(s)$ tends to a_n, from which it follows that $a_n = 0$, a contradiction. $\square$

Definition 4.13 We call the half-plane of Corollary 4.11 the *half-plane of convergence* of f (by abuse of terminology, $\emptyset$ and $\mathbf{C}$ are considered open half-planes). If the half-plane of convergence is given by $\Re(s) > \rho$, we say that $\rho = \rho(f)$ is the *abscissa of convergence* of f. (The cases $\emptyset$ and $\mathbf{C}$ correspond respectively to $\rho = +\infty$ and $\rho = -\infty$.) The *abscissa of absolute convergence* of f is the abcissa of convergence of $\sum |a_n| e^{-\lambda_n s}$, denoted $\rho^+(f)$.

Exercise 4.14 Let f be a formal Dirichlet series. Show that if $\rho(f) < +\infty$, then $\rho^+(f) - \rho(f) \leq 1$. (Use Proposition 1.1.)

The following theorem is a sort of converse to Theorem 4.10.

Theorem 4.15 (Landau, [75, pp. 536–537]) *Suppose that $a_n \geq 0$ for all $n \geq 1$, that f converges for $\Re(s) > \rho$, and that f extends analytically to a holomorphic function in a neighbourhood of $s = \rho$. Then there exists an $\varepsilon > 0$ such that f converges for $\Re(s) > \rho - \varepsilon$.*

Remark 4.16 This means that if f has coefficients ≥ 0, its domain of convergence is limited by a singularity of f lying on the real axis.

Proof Up to replacing s by $s - \rho$, we may suppose that $\rho = 0$. Since f is holomorphic both at $\Re(s) > 0$ and in a neighbourhood of 0, it is holomorphic in a disc $|s - 1| \leq 1 + \varepsilon$ where $\varepsilon > 0$. In particular, its Taylor series converges in this disc.

Using Lemma 1.3, we have:

$$f^{(p)}(s) = \sum_{n \geq 1} a_n(-\lambda_n)^p e^{-\lambda_n s} \text{ if } \Re(s) > 0,$$

whence

$$(-1)^p f^{(p)}(1) = \sum_{n \geq 1} \lambda_n^p a_n e^{-\lambda_n}.$$

The Taylor series in question is expressed, for $|s - 1| \leq 1 + \varepsilon$, as:

$$f(s) = \sum_{p=0}^{+\infty} \frac{1}{p!}(s - 1)^p f^{(p)}(1).$$

In particular, at $s = -\varepsilon$, we have:

$$f(-\varepsilon) = \sum_{p=0}^{+\infty} \frac{1}{p!}(1 + \varepsilon)^p (-1)^p f^{(p)}(1),$$

where the series converges.

But $(-1)^p f^{(p)}(1) = \sum_{n \geq 1} \lambda_n^p a_n e^{-\lambda_n}$ is a convergent series with terms ≥ 0. Consequently, the double series with positive terms

$$f(-\varepsilon) = \sum_{p,n} a_n \frac{1}{p!}(1 + \varepsilon)^p \lambda_n^p e^{-\lambda_n}$$

converges. We may thus re-group its terms and write it in the form:

$$f(-\varepsilon) = \sum_{n \geq 1} a_n e^{-\lambda_n} \sum_{p=0}^{\infty} \frac{1}{p!} (1+\varepsilon)^p \lambda_n^p$$

$$= \sum_{n \geq 1} a_n e^{-\lambda_n} e^{\lambda_n (1+\varepsilon)}$$

$$= \sum_{n \geq 1} a_n e^{\lambda_n \varepsilon}.$$

This shows that the given Dirichlet series converges for $s = -\varepsilon$, and thus also for $\Re(s) > -\varepsilon$ by Theorem 4.10. $\qquad\square$

Remarks 4.17

(1) As seen above, these results apply just as well to Dirichlet series as to power series.

(2) We have $\rho^+ \geq \rho$, and in general $\rho^+ > \rho$ for Dirichlet series, contrary to the case of power series (cf. Lemma 4.19, but see also Exercise 4.14).

Theorem 4.18 *Let* $f(s) = \sum \frac{a_n}{n^s}$ *be a Dirichlet series.*

(1) *Suppose that* $a_n = O((\log n)^{\alpha})$ *for some* $\alpha > 0$. *Then* $\rho^+(f) \leq 1$.

(2) *Suppose there exists* $\alpha > 0$ *and* $M > 0$ *such that the partial sums* $\sum_{m \leq n \leq m'} a_n$ *satisfy* $\left| \sum_{m \leq n \leq m'} a_n \right| \leq M (\log m)^{\alpha}$ *for all* $m \leq m'$. *Then* $\rho(f) \leq 0$.

Proof (1) Suppose there exists $M > 0$ such that we have $|a_n| \leq M(\log n)^{\alpha}$ for all $n \geq 1$. Then, $\sum \frac{|a_n|}{n^s} \leq M \sum_{n \geq 1} \frac{(\log n)^{\alpha}}{n^{\sigma}}$ and it is known that $\sum \frac{(\log n)^{\alpha}}{n^{\sigma}}$ converges for $\sigma > 1$.

For (2), let s be such that $\Re(s) > 0$: let us show that $f(s)$ converges. Theorem 4.10 even allows us to suppose that $s \in \mathbf{R}$. By applying Lemma 1.17, we get that:

$$\left| \sum_{m \leq n \leq m'} \frac{a_n}{n^s} \right| \leq \frac{M (\log n)^{\alpha}}{m^s}$$

which tends to 0 as m tends to infinity. $\qquad\square$

4.2.2 Convergence of Dirichlet L-functions

Lemma 4.19 *Let* χ *be a nontrivial Dirichlet character and let* $f = L(\chi, s)$. *Then* $\rho^+(f) = 1$ *and* $\rho(f) = 0$.

Proof Since we have $|\chi(n)| \le 1$ for all $n \ge 1$, the inequality $\rho^+(f) \le 1$ follows from Theorem 4.18 a). But there exists $m > 0$ such that $\chi(n) = 1$ for all $n \equiv 1 \pmod{m}$; as the series $\sum_{n \equiv 1 \pmod{m}} 1/n$ diverges, we have equality.

The inequality $\rho(f) \le 0$ follows from Theorem 4.18 b) by noting that the partial sums $\sum_{n=m_0}^{m_1} \chi(n)$ are bounded due to the "orthogonality" relation $\sum_{n=0}^{m-1} \chi(n) = 0$. (We recall here the proof: set $x = \sum_{n=0}^{m-1} \chi(n)$. For all n prime to m, we have $\chi(n)x = x$, and thus $x = 0$ since $\chi \ne 1$.) But for $\Re(s) \le 0$ the general term of $L(\chi, s)$ does not tend to 0, so $\rho(f) = 0$. $\qquad\square$

Theorem 4.20 *Let χ be a nontrivial Dirichlet character. Then $L(\chi, 1) \ne 0$.*

Corollary 4.21 *For $K = \mathbf{Q}(\mu_m)$, the function $\zeta_K(s)$ has a simple pole at $s = 1$.*

Theorem 4.20 is due to Gustav Lejeune-Dirichlet ([37], 1837). The proof below is essentially the one given in [Ser3, ch. VI, § 3.4].

Proof of theorem 4.20 Given Lemma 4.19, Proposition 1.15, Proposition 4.7, and Theorem 4.9, Theorem 4.20 is equivalent to Corollary 4.21. If this corollary is false, then $\zeta_K(s)$ is holomorphic for $\Re(s) > 0$. Since this is a Dirichlet series with positive real coefficients, Theorem 4.15 implies that it converges in this domain. We show however that this is absurd: if $\zeta_K(s) = \sum a_n n^{-s}$, then we have $a_n \ne 0$ if there is some ideal $\mathfrak{A} \subset O_K$ of norm n, which certainly holds when n is of the form $r^{[K:\mathbf{Q}]} = r^{\varphi(m)}$ $(N_{K/\mathbf{Q}}(rO_K) = r^{[K:\mathbf{Q}]}\mathbf{Z})$. This shows that the formal Dirichlet series $\zeta_K(s)$ is bounded below by $\zeta(\varphi(m)s)$, which diverges for $s = 1/\varphi(m)$ (Proposition 1.1). $\qquad\square$

Remarks 4.22

(1) To prove Theorem 4.20, we only need to use Proposition 4.8 which is somewhat more elementary than Theorem 4.9: this is what Serre does in [Ser3].

(2) Not having Landau's theorem at his disposal, Dirichlet proceeded differently. He distinguished two cases: χ real (that is to say, $\chi^2 = 1$) and χ non-real. In the second case, $L(\chi, 1) = 0 \Rightarrow L(\bar{\chi}, 1) = 0$ which would imply that $\zeta_K(s) \to 0$ as $s \to 1$, and yet we easily see that $\zeta_K(s) > 1$ for s real and > 1. In the first case, we have $\chi(n) = \left(\frac{d}{n}\right)$ (the Jacobi symbol) for a certain square-free d. Dirichlet calculated explicitly that

$$L(\chi, 1) = \begin{cases} \dfrac{2\pi h(d)}{w\sqrt{d}} & \text{if } d < 0, \\[3mm] \dfrac{\log(\varepsilon)h(d)}{\sqrt{d}} & \text{if } d > 0, \end{cases}$$

where $h(d)$ (resp. w) is the number of ideal classes (resp. roots of unity) of $\mathbf{Q}(\sqrt{d})$ and ε is the fundamental unit > 1 when $d > 0$. This is a special case of the analytic formula for the class number, from § 1.10.

See Davenport [Dav, ch. 6] for the details of this calculation and its advantages, as well as Stark [122] for a more elementary proof.

4.2.3 Application: the theorem on arithmetic progressions

We introduce the following Dirichlet series:

$$g_a(s) = \sum_{\substack{p \in \mathcal{P} \\ p \equiv a \pmod{m}}} \frac{1}{p^s}$$

where a is an integer coprime to m;

$$f_\chi(s) = \sum_{p \in \mathcal{P}} \frac{\chi(p)}{p^s}$$

where χ is a Dirichlet character modulo m.

An elementary calculation gives

Lemma 4.23 ("Fourier transform") *We have*

$$g_a(s) = \frac{1}{\varphi(m)} \sum_\chi \frac{f_\chi(s)}{\chi(a)},$$

where the sum is extended to all characters χ of $(\mathbf{Z}/m)^*$. □

Proposition 4.24 *The function $|\log L(\chi, s) - f_\chi(s)|$ is bounded as $s \to 1$ for all Dirichlet characters χ modulo m.*

Proof Proposition 4.6 gives

$$\log L(\chi, s) = \sum_p \log \frac{1}{1 - \chi(p)p^{-s}} = \sum_{n,p} \chi(p)^n / np^{ns}$$

and this equality holds for all $s \in \mathbf{C}$ such that $\Re(s) > 1$.

Then $\log L(\chi, s) = f_\chi(s) + F_\chi(s)$ where $F_\chi(s) = \sum_{p,n \geq 2} \frac{\chi(p)^n}{np^{ns}}$, and we easily see that $F_\chi(s)$ remains bounded when $s \to 1$. $\qquad \square$

Theorem 4.25 (Dirichlet [37]) *For any integer a coprime to m, we have*

$$g_a(s) \underset{s \to 1}{\sim} \frac{1}{\varphi(m)} \log \frac{1}{s-1}.$$

Proof Given Proposition 4.24, Proposition 1.15 implies that $f_1(s) \underset{s \to 1}{\sim} \log \frac{1}{s-1}$ whereas Theorem 4.20 implies that $f_\chi(s)$ remains bounded as $s \to 1$ for $\chi \neq 1$. We thus conclude by Lemma 4.23. $\qquad \square$

Corollary 4.26 *For every integer a coprime to m, there exists infinitely many prime numbers $p \equiv a \pmod{m}$.* $\qquad \square$

It was in order to prove this corollary that Dirichlet introduced his *L*-series.

4.2.4 Natural density and analytic density

Definition 4.27 Let T be a subset of the set $\mathcal{P}$ of prime numbers. We say that T has *natural density k* if the ratio

$$\frac{|\{p \in T \mid p < n\}|}{|\{p \in \mathcal{P} \mid p < n\}|}$$

tends to k as $n \longrightarrow \infty$.

Here is a second, less naïve, definition of density:

Definition 4.28 Let T be a subset of $\mathcal{P}$ and $k \in \mathbf{R}$. We say that T has *analytic density k* if the ratio

$$\frac{\displaystyle\sum_{p \in T} 1/p^s}{\displaystyle\sum_{p \in \mathcal{P}} 1/p^s}$$

tends to k as s tends to 1 in the interval $(1, +\infty)$.

Note that we necessarily have $0 \leq k \leq 1$ for these two notions and that a finite set has zero density: for the analytic density, we saw this during the proof of Theorem 4.25. The latter thus states that for $(a, m) = 1$ the set

$$\mathcal{P}_{a \pmod m} = \{p \in \mathcal{P} \mid p \equiv a \pmod m\}$$

has analytic density $\frac{1}{\varphi(m)}$, where φ is the Euler totient function.

If a set of prime numbers has a natural density, then it has an analytic density and these two densities coincide (Exercise 4.29). One may show that $\mathcal{P}_{a \pmod m}$ has a natural density (which is thus equal to $\frac{1}{\varphi(m)}$). Conversely, there exist $T \subset \mathcal{P}$ with a well-defined analytic density, but without a natural density (Bombieri, cf. [Ell, ch. 7, Exercises 14 and 15]).

4.2.5 Other applications of Dirichlet's theorem

Theorem 4.20 has important arithmetic applications to *analytic class number formulas* (for quadratic fields and cyclotomic fields $\mathbf{Q}(\mu_p)$ with p prime): see Remark 4.22 (2) and Borevič–Šafarevič [BS, ch. VI].

Exercise 4.29 (natural density and analytic density) Let T be as in Definition 4.28. We write

$$\pi_T(n) = |\{p \in T \mid p < n\}|, \quad \pi(n) = \pi_{\mathcal{P}}(n) \quad (n \in \mathbf{N})$$

$$F_T(s) = \sum_{p \in T} 1/p^s, \quad F(s) = F_{\mathcal{P}}(s) \quad (s > 1).$$

(a) Check the identity

$$F_T(s) = \sum_{n=1}^{\infty} \frac{\pi_T(n+1) - \pi_T(n)}{n^s}.$$

(b) From this, deduce the identity

$$F_T(s) = \sum_{n=1}^{\infty} \pi_T(n+1) \left(\frac{1}{n^s} - \frac{1}{(n+1)^s} \right).$$

(c) Suppose that T has natural density δ. Let $\varepsilon > 0$: then there exists $n_0 \gg 0$ such that $|\pi_T(n) - \delta \pi(n)| < \varepsilon \pi(n)$ whenever $n > n_0$. By considering $F_T(s) - \delta F(s)$, show that T has analytic density δ. (Separate the series into two parts, $\sum_{n=1}^{n_0-1}$ and $\sum_{n=n_0}^{\infty}$, and use the fact that $\lim_{s \to 1} F(s) = +\infty$, cf. the proof of Theorem 4.25.)

Exercise 4.30 Let $a \in \mathbf{Z} - \{0\}$.

(a) If p is a prime number not dividing $2a$, show that we have $a^{\frac{p-1}{2}} \equiv \pm 1$ (mod p). We write $\left(\frac{a}{p}\right)$ for the element of $\{\pm 1\} \subset \mathbf{Z}$ which gets sent to $a^{\frac{p-1}{2}}$ modulo p: this is the *Legendre symbol of a modulo p*.

(b) Show that for odd p there exists a b such that $\left(\frac{b}{p}\right) = -1$.

For the remainder of the exercise we admit *the law of quadratic reciprocity*

$$\left(\frac{p}{q}\right)\left(\frac{q}{p}\right) = (-1)^{\frac{p-1}{2} \cdot \frac{q-1}{2}}$$

for two distinct odd primes p, q, with complementary formulae

$$\left(\frac{-1}{p}\right) = (-1)^{\frac{p-1}{2}}, \qquad \left(\frac{2}{p}\right) = (-1)^{\frac{p^2-1}{8}}.$$

(c) Suppose that a is an odd prime, and let m be an integer prime to $2a$: using the law of quadratic reciprocity, give an expression of the product $\prod_{l|m} \left(\frac{a}{l}\right)^{v_l(m)}$ in terms of $\left(\frac{m}{a}\right)$.

(d) By using the complementary formulas, give an expression of the same product when $a = -1$ and $a = 2$.

(e) Show that there exists a unique Dirichlet character χ_a modulo $4a$ such that $\chi_a(p) = \left(\frac{a}{p}\right)$ for all prime numbers p not dividing $2a$. (Reduce to the case where a is prime or $a = -1$, and use (c), (d).)

(f) If a is prime or $a = -1$, show that $\chi_a \neq 1$.

(g) If a is not a square, show that $\chi_a \neq 1$. (Write $a = l^n a_1$ with l prime, n odd, and $(a_1, l) = 1$; use (f) and the Chinese remainder theorem.)

(h) Suppose that a is a square modulo p for all but finitely many $p \in \mathcal{P}$. Show that a is a square. (Use (g) and the theorem on arithmetic progressions.)

4.3 First generalisations: Hecke L-functions

4.3.1 Restricted direct products

Definition 4.31 Let $(G_i)_{i \in I}$ be a family of sets, and for all $i \in I$, let U_i be a subset of G_i. We write

$$\prod_{i \in I} (G_i, U_i) = \left\{ (g_i) \in \prod_i G_i \mid \{j \in I \mid g_j \notin U_j\} \text{ is finite} \right\}.$$

This is the *restricted direct product of the G_i relative to the U_i*.

We clearly have

$$\prod_{i\in I}(G_i, U_i) = \varinjlim_{J\subset I \text{ finite}} \prod_{i\in J} G_i \times \prod_{i\notin J} U_i.$$

4.3.2 Places of global fields

Let K be a global field, i.e. a number field or a function field in one variable over a finite field. Write Σ_K for the set of places of K (equivalence classes of nontrivial absolute values). We also write Σ_K^f for the set of nonarchimedean places and Σ_K^∞ for the set of archimedean places. If char $K = 0$, then the $v \in \Sigma_K^f$ correspond bijectively with the prime ideals of O_K and the $v \in \Sigma_K^\infty$ with embeddings $K \hookrightarrow \mathbf{C}$ up to complex conjugation; if char $K > 0$, then we have $\Sigma_K^\infty = \emptyset$ and $\Sigma_K = \Sigma_K^f$ is in bijection with the closed points of a curve C, the unique smooth projective model of K over its field of constants.

For $v \in \Sigma_K$, we write K_v for the completion of K with respect to v. Thus:

(i) If K is a number field then K_v is a p-adic field if v is nonarchimedean, corresponding to an ideal $\mathfrak{p} \subset O_K$ over p, and $K_v = \mathbf{R}$ or $\mathbf{C}$ if v is archimedean.

(ii) If K is a function field with field of constants $k = \mathbf{F}_q$, we have $K_v \simeq k_v((t))$ where k_v is a finite extension of k (the residue field of C at v, if C is the projective smooth model of K/k).

4.3.3 Adeles and Ideles

Let K be a global field.

Definition 4.32

a) The *adele ring of K* is the ring

$$\mathbb{A}_K = \prod_{v\in\Sigma_K}(K_v, O_v)$$

where O_v is the valuation ring of v if v is nonarchimedean and $O_v = K_v$ if v is archimedean.

b) The *idele group* of K is the group of units $\mathbb{I}_K$ of $\mathbb{A}_K$:

$$\mathbb{I}_K = \prod_{v\in\Sigma_K}(K_v^*, O_v^*).$$

c) If K is a number field, we also write

$$\mathbb{A}_K^f = \prod_{v \in \Sigma_K^f} (K_v, O_v)$$

(the finite adeles) and $\mathbb{I}_K^f$ for the corresponding ideles.

Remark 4.33 Here is a different description of $\mathbb{A}_K$ if char $K = 0$: we have $\mathbb{A}_K = \mathbb{A}_\mathbf{Z} \otimes_\mathbf{Z} K$, with

$$\mathbb{A}_\mathbf{Z} = \mathbf{R} \times \prod_p \mathbf{Z}_p.$$

In characteristic p, we may do the same thing by using the ring

$$\mathbb{A} = \prod_v O_v$$

where v runs over all places of K.

Ideles were introduced by Claude Chevalley in 1936 [19]. Adeles were introduced a little later by Artin and Whaples under the name of *valuation vectors* [11], a name replaced by "Adele" by Weil. The goal was to reformulate class field theory in a more conceptual form. An older notion of "rings of distributions" (without completions) is due to Weil.

4.3.4 The topology of adeles and ideles

References: Cassels–Fröhlich [CF, ch. II], Lang [Lan2, ch. VII]. See also Iwasawa [56].

We start from a family $(G_i, U_i)_{i \in I}$ of *topological spaces* as in Section 4.3.1. Suppose that the G_i are locally compact, and the U_i are open in G_i for all i; suppose furthermore that U_i is compact for all but finitely many i. Then the products $\prod_{i \in J} G_i \times \prod_{i \notin J} U_i$ are locally compact, and their inductive limit $\prod_{i \in I}(G_i, U_i)$ is *locally compact* for the inductive limit topology.

This applies to $\mathbb{A}_K$, $\mathbb{I}_K$, $\mathbb{A}_K^f$ and $\mathbb{I}_K^f$.

Warning 4.34 The inclusion $\mathbb{I}_K \hookrightarrow \mathbb{A}_K$ is continuous, but the topology of $\mathbb{I}_K$ is stronger than the topology induced from that of $\mathbb{A}_K$ (for which $x \mapsto x^{-1}$ is not continuous). The topology of $\mathbb{I}_K$ is induced by the embedding

$$\mathbb{I}_K \to \mathbb{A}_K \times \mathbb{A}_K,$$
$$x \mapsto (x, x^{-1}),$$

with closed image.

We have the diagonal embedding $K \hookrightarrow \mathbb{A}_K$, and

Theorem 4.35

(1) *K is discrete in $\mathbb{A}_K$; the quotient $\mathbb{A}_K / K$ is compact.*
(2) **Strong approximation theorem.** *Let $v_0 \in \Sigma_K$. Then the diagonal embedding*

$$K \hookrightarrow \prod_{v \neq v_0} (K_v, O_v)$$

has dense image.

Sketch For (1), let $K_0 = \mathbf{Q}$ or $\mathbf{F}_q(t)$. The choice of a basis of K over K_0 provides an isomorphism $\mathbb{A}_K \simeq \mathbb{A}_{K_0}^{[K:K_0]}$ that is compatible with the diagonal embeddings. We thus reduce to the case $K = K_0$, and this follows from a simple calculation [CF, p. 65]. For (2), we use an adelic form of Minkowski's theorem on lattice points in $\mathbf{R}^N$, cf. [CF, ch. II, § 15].   □

For the case of ideles, we must introduce the *norm*

$$\| \ \| : \mathbb{I}_K \to \mathbf{R}^{+*}, \tag{4.3.1}$$

$$\|(a_v)_{v \in \Sigma_K}\| = \prod_v \|a_v\|_v$$

where $\| \ \|_v$ is the normalised "absolute value" on K_v for all v. We recall:

$K_v = \mathbf{R}$: $\|a\|_v = |a|$.
$K_v = \mathbf{C}$: $\|a\|_v = |a|^2$.
K_v **is nonarchimedean:** $\|a\|_v = N(v)^{-v(a)}$, where $N(v)$ is the cardinality of the residue field of K_v and $v(a)$ is the normalised valuation of a (i.e. $v(\pi) = 1$ if π is a uniformiser).

Theorem 4.36 *The diagonal embedding*

$$K^* \hookrightarrow \mathbb{I}_K$$

has discrete image. This image is contained in $\mathbb{I}_K^0 = \mathrm{Ker}(\| \ \|)$, and $\mathbb{I}_K^0 / K^*$ is compact.

Proof The first assertion follows easily from Theorem 4.35(1). The fact that $\|a\| = 1$ for all $a \in K^*$ is called the *product formula*. To show it, we check that if $K_0 \subset K$ then we have the equality $\|a\| = \|N_{K/K_0}(a)\|$ (this extends to $\mathbb{I}_K$ and $\mathbb{I}_{K_0}$ for a suitable definition of the norm). We are thus brought to the case $K = \mathbf{Q}$ or $\mathbf{F}_q(t)$; in both of these cases we have a ring of integers that is a PID, and the proof is now easy. (For another proof, see [126, th. 4.3.1].)

For the compactness of the quotient, we use Theorem 4.35(2), showing first that the topology of $\mathbb{I}_K^0$ is induced by that of $\mathbb{A}_K$, cf. 4.34 [CF, ch. II, § 16]. □

Definition 4.37 The quotient $C_K = \mathbb{I}_K/K^*$ is called the *idele class group*.

We thus have an exact sequence

$$0 \to C_K^0 \to C_K \xrightarrow{\| \| } \mathbf{R}^{+*} \to 0$$

where C_K^0 is the group of idele classes of norm 1.

4.3.5 Connection with "classical" algebraic number theory

Ideles were introduced in order to describe generalised class groups:

If A is a Dedekind ring, the ideal class group $\mathrm{Cl}(A)$ is the quotient of the group of fractional ideals by the group of principal fractional ideals: it is identified with the Picard group $\mathrm{Pic}(\mathrm{Spec}\, A)$.

We generalise this definition as follows. Let K be a global field.

Definition 4.38 We write I_K for the group of divisors of K:

(i) If char $K = 0$, then this is the group of fractional ideals of the ring of integers O_K.

(ii) If char $K = p > 0$, we have $K = \mathbf{F}_q(C)$ for a smooth projective geometrically connected curve C, where q is a power of p. We write $I_K = \mathrm{Div}(C) = Z_0(C)$.

We have a surjective homomorphism

$$\varphi : \mathbb{I}_K \to I_K,$$
$$(a_v) \mapsto \sum_{v \in \Sigma_K^f} v(a_v)v,$$

which is continuous (for the discrete topology on I_K) and such that $\varphi(f) = (f)$ (the divisor of f) for all $f \in K^*$.

Lemma 4.39 *If* char $K = 0$, *the restriction of* φ *to* $\mathbb{I}_K^0$ *is surjective.*

Proof If $\mathfrak{p}$ is a prime ideal of O_K, then the idele (a_v) with $a_v = 1$ for $v \neq \mathfrak{p}$, $v_\mathfrak{p}(a_\mathfrak{p}) = 1$, $\|a_v\|_v = \|a_\mathfrak{p}\|_\mathfrak{p}^{-1}$ for an archimedean place v and $a_v = 1$ for the other archimedean places, is in $\mathbb{I}_K^0$ and is sent to $\mathfrak{p}$. □

Definition 4.40 A *modulus* is a formal linear combination

$$\mathfrak{m} = \sum_{v \in \Sigma_K} n_v v$$

where $n_v \in \mathbf{N}$ is zero for all but finitely many v, and $n_v = 0$ or 1 when v is archimedean. The set $S = \{v \mid n_v \neq 0\}$ is the *support of* $\mathfrak{m}$.

Definition 4.41 Let $\mathfrak{m} = \sum n_v v$ be a modulus of S.

a) We write I_K^S for the subgroup of I_K formed from those $\mathfrak{a}$ prime to S: $v(\mathfrak{a}) = 0$ if v is a finite place of S.
b) Let $f \in K^*$. We write $f \equiv 1 \pmod{*\mathfrak{m}}$ if
   (i) $v(f - 1) \geq n_v$ whenever v is finite and $n_v > 0$;
   (ii) $f > 0$ if v is real and $n_v = 1$.
c) We write

$$K^{\mathfrak{m}} = \{f \in K^* \mid f \equiv 1 \pmod{*\mathfrak{m}}\}$$

and $\mathrm{Cl}^{\mathfrak{m}}(K)$ for the quotient $I_K^S/(K^{\mathfrak{m}})$. This is the *generalised class group* relative to the modulus $\mathfrak{m}$. (It is often called the *ray class field*.)

Examples 4.42

(1) $\mathfrak{m} = 0$: We recover the class group (of O_K in characteristic zero and of C in characteristic p).
(2) $K = \mathbf{Q}$, $\mathfrak{m} = m$. We get an isomorphism $(\mathbf{Z}/m)^* \xrightarrow{\sim} \mathrm{Cl}^m(\mathbf{Q})$.

Returning to ideles, if $\mathfrak{m}$ is a modulus, we write

$$\mathbb{I}_K^{\mathfrak{m}} = \{a \in \mathbb{I}_K \mid a \equiv 1 \pmod{*\mathfrak{m}}\}$$

(a direct generalisation of Definition 4.41), so that $K^{\mathfrak{m}} = K^* \cap \mathbb{I}_K^{\mathfrak{m}}$. The homomorphism φ from Definition 4.38 sends $\mathbb{I}_K^{\mathfrak{m}}$ into I_K^S. Moreover:

Proposition 4.43 *We have a diagram*

$$\mathbb{I}_K^{\mathfrak{m}}/K^{\mathfrak{m}} \xrightarrow{i} C_K$$
$$\bar{\varphi} \downarrow$$
$$\mathrm{Cl}^m(K)$$

where i, induced by the inclusion $\mathbb{I}_K^{\mathfrak{m}} \subset \mathbb{I}_K$, is bijective and $\bar{\varphi}$, induced by φ, is surjective and continuous (for the discrete topology on $\mathrm{Cl}^{\mathfrak{m}}(K)$). If char $K = 0$, then $\mathrm{Cl}^{\mathfrak{m}}(K)$ is finite.

Sketch Surjectivity of i follows easily from the strong approximation theorem, and even from less. For finiteness, Lemma 4.39 shows that the restriction of $\bar\varphi$ to $(\mathbb{I}_K^m \cap \mathbb{I}_K^0)/K^m$ is surjective, but this implies that $\mathrm{Cl}^m(K)$, the discrete quotient of a compact group, is finite. $\qquad\square$

Remark 4.44 The finiteness of $\mathrm{Cl}^m(K)$ is false if $\mathrm{char}\,K > 0$: on the one hand, Lemma 4.39 is false in this case, and we even have a commutative diagram

$$
\begin{array}{ccc}
\mathbb{I}_K^m/K^m & \xrightarrow{\ \|\,\|\ } & \mathbf{R}^* \\
\scriptstyle\bar\varphi \downarrow & & \uparrow \scriptstyle\rho \\
\mathrm{Cl}^m(K) & \xrightarrow{\ \deg\ } & \mathbf{Z}
\end{array}
$$

where $\rho(1) = q^{-1}$ if $\mathbf{F}_q$ is the field of constants of K, and where the homomorphism deg is surjective. It follows that $\mathrm{Cl}^m(K)^0 := \mathrm{Ker}(\deg)$ is finite, which is the correct statement.

To have a perfect analogy between characteristic zero and characteristic p, we would need to introduce groups of compactified divisors (equipped with metrics at infinity) in characteristic zero, cf. Szpiro [124, § 1]. This idea goes back to Weil, and gave rise to Arakelov theory.

4.3.6 Hecke characters or Größencharaktere: the idelic version

Definition 4.45 A *quasicharacter* of a locally compact group G is a continuous homomorphism $\chi : G \to \mathbf{C}^*$. This is a *character* if it takes values in the unit circle $S^1 \subset \mathbf{C}^*$.

Definition 4.46 A quasicharacter $\chi : \mathbb{I}_K \to \mathbf{C}^*$ is *unramified at* $v \in \Sigma_K^f$ if $\chi(O_v^*) = 1$.

Definition 4.47 A *Hecke character* is a quasicharacter of C_K.

Example 4.48 The norm $\|\ \|$ (4.3.1) is a Hecke character, which we call *principal*.

Lemma 4.49 *Let χ be a Hecke character. We may write*

$$\chi = \prod_v \chi_v$$

where $\chi_v = \chi_{|K_v^*}$; furthermore, χ_v is unramified, that is to say $\chi_v(O_v^*) = 1$, for all but finitely many $v \in \Sigma_K^f$.

Proof The assertion of being "unramified" follows from the continuity of χ.
$\square$

The following lemma allows us to recover quasicharacters from characters:

Lemma 4.50 *Let χ be a Hecke character. Then there exists a unique $\sigma \in \mathbf{R}$ such that the Hecke character*

$$a \mapsto \chi(a)\|a\|^{-\sigma}$$

takes values in S^1 (i.e. is a "true" character). We call σ the exponent *of χ. If χ has exponent 0, we say that it is* unitary.

Proof The restriction of $|\chi|$ to the compact group C_K^0 is trivial, since $\mathbf{R}^{+*}$ has no nontrivial compact subgroup. So $|\chi|$ factors through $\| \, \|$. But every continuous endomorphism of $\mathbf{R}^{+*}$ is of the form $x \mapsto x^\sigma$ for some $\sigma \in \mathbf{R}$, as one sees by passing to the logarithm.
$\square$

4.3.7 Passage to admissible characters

This exposition is taken from Tate's account in [CF, ch. VII, §§ 3,4].

Definition 4.51 Let $S \subset \Sigma_K$ be a finite set containing Σ_K^∞; write I_K^S for the set of divisors prime to (the set of finite places of) S. If $f \in K^*$, we write $(f)^S$ for the part of the divisor of f that belongs to I^S, that is, $(f)^S = \sum_{v \in \Sigma_K^f - S} v(f)v$.

Definition 4.52 An *admissible pair* is a pair (S, φ) where S is as above, and φ is a homomorphism from I_K^S to a commutative topological group G satisfying:

For every neighbourhood U of the identity $1 \in G$, there exists $\varepsilon > 0$ such that $\varphi((f)^S) \in U$ whenever $|f - 1|_v < \varepsilon$ for all $v \in S$.

Proposition 4.53 ([CF, ch. VII, prop. 4.1]) *Via the isomorphism i from Proposition 4.43, any admissible pair (S, φ) induces a unique continuous homomorphism $\psi : C_K \to G$ such that $\psi(a) = \varphi((a)^S)$ for all $a \in \mathbb{I}_K^S$, where*

$$\mathbb{I}_K^S = \{(a_v) \in \mathbb{I}_K \mid |a_v| = 1 \text{ for } v \in S\}$$

and $(a)^S$ is defined as in Definition 4.51.
Conversely, if there exists a neighbourhood U of $1 \in G$ such that the only subgroup of G contained in U is equal to $\{1\}$, then all continuous homomorphisms $\psi : C_K \to G$ arise from an admissible pair.

We note that the admissible pair (S, φ) in Proposition 4.53 is not uniquely determined as a function of ψ: this corresponds to the ambiguity arising from non-primitive Dirichlet characters, and leads to a generalisation of the conductor:

Definition 4.54 Let $\chi : C_K \to \mathbf{C}^*$ be a Hecke character. We define the *conductor* of χ to be the divisor $\mathfrak{f}(\chi) = \sum_{v \in \Sigma_K^f} n_v v$ where n_v is the smallest integer n such that $\chi_v(1 + \mathfrak{p}_v^n) = 1$, where $\mathfrak{p}_v$ is the maximal ideal of O_v ($n_v = 0$ if $\chi_v(O_v^*) = 1$).

We also have the notion of *infinity-type*: See [CF, ch. VIII, § 1], and Jerry Shurman's notes [119] for a more detailed description of the situation. Briefly, by applying Proposition 4.53 to a Hecke character χ, we may see χ as a homomorphism

$$\chi : I_K^S \to \mathbf{C}^*$$

such that, if $f \equiv 1 \pmod{* \mathfrak{f}(\chi)}$, then $\chi((f))$ is determined by the infinity-type of χ: this was Hecke's original definition [50]. We only give two examples:

Examples 4.55

(1) Continuing from Example 4.42, we see that Dirichlet characters are particular instances of unitary Hecke characters.
(2) For any global field K and any $s \in \mathbf{C}$, the function $\mathfrak{A} \mapsto N(\mathfrak{A})^{-s}$ defines a Hecke character (which corresponds to the function $\| \ \|^s : C_K \to \mathbf{C}^*$).

4.3.8 Hecke L-functions

Definition 4.56 Let χ be a Hecke character with conductor $\mathfrak{f}$. We let

$$L(\chi, s) = \sum_{\mathfrak{a} \in I_K^+} \frac{\chi(\mathfrak{a})}{N(\mathfrak{a})^s},$$

where I_K^+ denotes the monoid of positive divisor of K and where $\chi(\mathfrak{a}) := 0$ if $(\mathfrak{a}, \mathfrak{f}) \neq 1$.

Example 4.57 For $\chi = 1$, we recover the Dedekind zeta function from § 1.10.

Proposition 4.58

(1) *We have the identity*

$$L(\chi \cdot \| \ \|^t, s) = L(\chi, s + t).$$

(2) *We have $\rho^+(L(\chi, s)) \le 1 + \sigma$, where σ is the exponent of χ (Lemma 4.50).*

(3) *We have the usual Euler product*

$$L(\chi, s) = \prod_{\mathfrak{p} \nmid \mathfrak{m}} \left(1 - \frac{\chi(\mathfrak{p})}{N(\mathfrak{p})^s} \right)^{-1}.$$

Proof (1) is clear (cf. Example 4.55 (2)). Then (2) follows from Theorem 4.18 (or the case of the Riemann zeta function) by reducing to the case of unitary χ. Finally, (3) can be proved as usual. □

Remark 4.59 The identity from Proposition 4.58 (1) can also be written as

$$L(\chi \cdot \|\ \|^t, s) = L(\chi \cdot \|\ \|^{t+s}, 1)$$

which allows us to view a Hecke L-function as a continuous function on the space of Hecke characters. This interpretation is the starting point of Tate's proof of the following theorem.

Theorem 4.60 (Hecke, 1920 [50]) *Let χ be a unitary Hecke character. Set*

$$\Lambda(\chi, s) = (|d_K| N(\mathfrak{f}(\chi)))^{s/2} \prod_{v \in \Sigma_K^{\mathbf{R}}} \Gamma_{\mathbf{R}}(s + n_v) \prod_{v \in \Sigma_K^{\mathbf{C}}} \Gamma_{\mathbf{C}}(s + \frac{|n_v|}{2}) L(\chi, s)$$

where $\Gamma_{\mathbf{R}}, \Gamma_{\mathbf{C}}$ have been defined in (1.10.1), d_K is the absolute discriminant of K if char $K = 0$, and $d_K = q^{2g-2}$ if $K = \mathbf{F}_q(C)$ where C is a smooth, projective, geometrically integral curve over $\mathbf{F}_q$ of genus g, and

v real: $n_v = 1$ if $\chi_v \ne 1$ and $n_v = 0$ if $\chi_v = 1$.
v complex: n_v is the integer n such that $\chi_v(z) = z^n$ for z of modulus 1. In particular, $n_v = 0$ if χ has finite order.

Then $\Lambda(\chi, s)$ admits a meromorphic extension to the whole complex plane and satisfies the functional equation

$$\Lambda(\bar{\chi}, 1 - s) = W(\chi)\Lambda(\chi, s)$$

where $(\bar{\chi})(x) := \overline{\chi(x)}$, and $W(\chi) \in \mathbf{C}^*$ has a canonical decomposition

$$W(\chi) = \prod_{v \in \Sigma_K} W(\chi_v)$$

in which $|W(\chi_v)| = 1$ and $W(\chi_v) = 1$ if χ_v is unramified (in general, $W(\chi_v)$ is a normalised Gauss sum).

If χ is of the form $\|\ \|^{it}$ for $t \in \mathbf{R}$, then $L(\chi, s)$ has a simple pole at $s = 1 + it$ and is holomorphic elsewhere. Otherwise, $L(\chi, s)$ is holomorphic in the whole complex plane.

Tate's proof [126] proceeds by doing Fourier analysis on the space of adeles; the key step is the Poisson formula, which is an analytic analogue of the Riemann–Roch theorem. This avoids Hecke's recourse to theta functions, generalising the proof of Theorem 1.18.

4.4 Second generalisation: Artin *L*-functions

For more details, see Martinet [82].

4.4.1 Review of ramification theory

Reference: any good book on algebraic number theory.

Let A be a Dedekind domain with fraction field K: if $\mathfrak{p}$ is a prime ideal of A, we write $\kappa(\mathfrak{p}) = A/\mathfrak{p}$ for its *residue field*.

Let L be a finite separable extension of K and let B be the integral closure of A in L: this is a Dedekind domain, finite as a module over A. If $\mathfrak{P}$ is a prime ideal of B, $\mathfrak{p} = A \cap \mathfrak{P}$ is a prime ideal of A: we say that $\mathfrak{P}$ lies *over* $\mathfrak{p}$ or that $\mathfrak{P}$ divides $\mathfrak{p}$ (notation: $\mathfrak{P} \mid \mathfrak{p}$). Given any $\mathfrak{p}$, we have

$$\mathfrak{p}B = \prod_{\mathfrak{P}\mid\mathfrak{p}} \mathfrak{P}^{e_{\mathfrak{P}}}$$

where $e_{\mathfrak{P}}$ is (defined to be) the *ramification index* of $\mathfrak{P}$. We have the formula

$$[L : K] = \sum_{\mathfrak{P}\mid\mathfrak{p}} e_{\mathfrak{P}} f_{\mathfrak{P}} \qquad (4.4.1)$$

where $f_{\mathfrak{P}} = [\kappa(\mathfrak{P}) : \kappa(\mathfrak{p})]$ is the *residue class degree* of $\mathfrak{P}$.

Suppose that L/K is Galois with Galois group G. Then all $\mathfrak{P} \mid \mathfrak{p}$ are conjugate under the action of G. We may write $e_{\mathfrak{P}} = e_{\mathfrak{p}}$, $f_{\mathfrak{P}} = f_{\mathfrak{p}}$, and then (4.4.1) reduces to

$$[L : K] = e_{\mathfrak{p}} f_{\mathfrak{p}} g_{\mathfrak{p}}, \qquad (4.4.2)$$

where $g_{\mathfrak{p}}$ is the number of prime ideals above $\mathfrak{p}$. Choose $\mathfrak{P}$ above $\mathfrak{p}$: we write

$$D_{\mathfrak{P}} = \{g \in G \mid g\mathfrak{P} = \mathfrak{P}\}$$

for the *decomposition group of* $\mathfrak{P}$. It acts on the residue field $\kappa(\mathfrak{P})$; the extension $\kappa(\mathfrak{P})/\kappa(\mathfrak{p})$ is Galois, and the induced homomorphism

$$D_{\mathfrak{P}} \to \text{Gal}(\kappa(\mathfrak{P})/\kappa(\mathfrak{p}))$$

is *surjective*. Its kernel $I_{\mathfrak{P}}$ is the *inertia group at* $\mathfrak{P}$. We have

$$(G : D_{\mathfrak{P}}) = g_{\mathfrak{p}}, \quad (D_{\mathfrak{P}} : I_{\mathfrak{P}}) = f_{\mathfrak{p}}, \quad |I_{\mathfrak{P}}| = e_{\mathfrak{p}}.$$

Let $\mathfrak{P}' \mid \mathfrak{p}$: there exists $g \in G$ such that $\mathfrak{P}' = g\mathfrak{P}$, and we have

$$D_{\mathfrak{P}'} = gD_{\mathfrak{P}}g^{-1}, \quad I_{\mathfrak{P}'} = gI_{\mathfrak{P}}g^{-1}.$$

4.4.2 The case of curves

Suppose that $K = k(C)$, where C is a smooth, projective, geometrically integral k-curve. By considering the local rings of C we obtain a local ramification theory, which we can then globalise.

Let L be a finite separable extension of K, with field of constants l. To simplify, suppose that $l = k$: we may always reduce to this case by replacing K by Kl.

Let C' be the smooth projective model of L. The morphism $f : C' \to C$ is finite and flat: it is a "ramified covering". The global information provided by this ramification is the *Riemann–Hurwitz–Hasse formula* [48]

$$2g' - 2 = [L : K](2g - 2) + \sum_{x \in C'_{(0)}} \delta_x$$

where g is the genus of C, g' is that of C', and δ_x is the valuation at x of the differential of f for all closed points x of C'. If f is tamely ramified at x, we have $\delta_x = e_x - 1$ where e_x is the ramification index of the prime ideal of $\mathcal{O}_{C',x}$ [Har2, ch. IV, prop. 2.2].

(When k is algebraically closed, this formula is proved in Hartshorne [Har2, ch. IV, cor. 2.4]: in the general case, one may either rewrite the proof or reduce to this case, noting that g, g' and δ_x are invariant under extension of scalars.)

Suppose that L/K is Galois with Galois group G: then the extension of fields of constants l/k is Galois with Galois group g, and the homomorphism $G \to g$ is *surjective*.

4.4.3 The Frobenius automorphisms

Let K be a global field, and let L/K be a Galois extension with Galois group G. Let v be a finite place of K and w a finite place of L dividing v. The residue field $\kappa(w)$ has a canonical automorphism: the *Frobenius automorphism* $\varphi_{w/v}$. If $\kappa(v) = \mathbf{F}_q$, we have

$$\varphi_{w/v}(x) = x^q$$

for $x \in \kappa(w)$.

Definition 4.61 The *Frobenius automorphism of G at w* is the element of D_w/I_w corresponding to $\varphi_{w/v}$: we keep the notation $\varphi_{w/v}$.

Suppose that v is unramified, so that $I_w = 1$ for all $w \mid v$. Then the set of $\varphi_{w/v}$ forms a conjugacy class of G, denoted φ_v.

4.4.4 Artin *L*-functions

Let $\rho : G \to GL(V)$ be a linear representation on a finite-dimensional **C**-vector space V.

Definition 4.62 (Artin, 1923–1930 [7, 9]) The *L-function of ρ* is

$$L(\rho, s) = \prod_{v \in \Sigma_K^f} L_v(\rho, s)$$

with

$$L_v(\rho, s) = \det(1 - \rho(\varphi_{w/v})N(v)^{-s} \mid V^{I_w})^{-1}.$$

We explain this definition: the quotient group D_w/I_w acts on V^{I_w}; the characteristic polynomial of $\rho(\varphi_{w/v})$ acting on this space is independent of the choice of w. Artin started by defining his *L*-functions up to ramified factors [7], and then understood the formula for the latter in [9].

Proposition 4.63 *Artin's L-functions converge absolutely for* $\Re(s) > 1$ *and have the following properties:*

(i) $L(\mathbf{1}, s) = \zeta_K(s)$; $L(\rho \oplus \rho', s) = L(\rho, s)L(\rho', s)$.
(ii) *Let* L'/K *be a Galois extension containing* L/K, *with Galois group* G'. *Denote by* $\tilde\rho$ *the representation of* G' *induced from* ρ *(by "inflation"). Then* $L(\tilde\rho, s) = L(\rho, s)$.
(iii) *Let* M/K *be a sub-extension of* L/K, *and let* $H = \mathrm{Gal}(L/M)$. *Let* θ *be a linear representation of* H. *Then* $L(\mathrm{Ind}_H^G \theta, s) = L(\theta, s)$.

Proof Absolute convergence can be shown by bounding from above by the Dedekind zeta function of K. The properties (i) and (ii) are trivial[1]; we show (iii) which is the most important.

Let $v \in \Sigma_K^f$: by abuse of notation, we write

$$L_v(\rho, s) = \det(1 - \rho(\varphi_v)N(v)^{-s} \mid V^{I_v})^{-1}$$

ignoring the mention (and choice of) w. Let $\Sigma(v)$ be the set of places of M over v. For precision, we write $I_{\tilde{w}/w}$ and $I_{\tilde{w}/v}$ for the corresponding inertia groups. Let V be the vector space underlying the representation of θ, and let W be the space underlying the representation of $\hat{\theta} := \mathrm{Ind}_H^G \theta$. We need only show that

$$\det(1 - \hat{\theta}(\varphi_v)N(v)^{-s} \mid W^{I_v})^{-1} = \prod_{w \in \Sigma(v)} \det(1 - \theta(\varphi_w)N(w)^{-s} \mid V^{I_w})^{-1}.$$

For $w \in \Sigma(v)$, set $f_w = f_{w/v} = [\kappa(w) : \kappa(v)]$. Since $N(w) = N(v)^{f_w}$, in view of (3.6.2) we must prove the identity

$$\sum_{n=1}^{\infty} \mathrm{Tr}(\hat{\theta}(\varphi_v^n) \mid W^{I_v}) \frac{t^n}{n} = \sum_{w \in \Sigma(v)} \sum_{n=1}^{\infty} \mathrm{Tr}(\theta(\varphi_w^n) \mid V^{I_w}) \frac{t^{n f_w}}{n} \qquad (4.4.3)$$

which is an amusing exercise (Exercise 4.64). □

Property (ii) allows us to speak of $L(\rho, s)$ for a complex (continuous) representation of the *absolute Galois group* of K, without choosing a finite Galois extension L/K.

Exercise 4.64

(a) Let G be a cyclic group, and let H be a subgroup of G of index f. Let $\theta : H \to GL(V)$ be a complex representation of H, let W be the representation space of $\hat{\theta} = \mathrm{Ind}_H^G \theta$ and let φ be a generator of H. Show that

$$\mathrm{Tr}\left(\hat{\theta}(\varphi^n) \mid W\right) = \begin{cases} 0 & \text{if } f \nmid n, \\ f \, \mathrm{Tr}\left(\theta(\varphi^{n/f}) \mid V\right) & \text{if } f \mid n. \end{cases}$$

(By decomposing θ into a direct sum of degree 1 representations [Ser1, 3.1, th. 9], observe that this representation is the restriction of a representation θ' of G; use the formula

$$\mathrm{Ind}_H^G \mathrm{Res}_H^G \theta' \simeq \theta' \otimes \mathrm{Ind}_H^G \mathbf{1}$$

---

[1] For (ii), the point is that if w' is a place of L' over w, the homomorphism $I_{w'} \to I_w$ is surjective.

[Ser1, 3.3, ex. 5], and then the formulas giving the character of a tensor product and those of the regular representation [Ser1, 2.1, prop. 2 and 2.4, prop. 5].)

(b) Let G be a finite group, H and D be subgroups of G and I a normal subgroup of D. Let θ be a complex representation of H and $\hat\theta = \mathrm{Ind}_H^G \theta$. Prove the identity

$$(\mathrm{Res}_D^G \hat\theta)^I \simeq \bigoplus_{s \in D\backslash G/H} \mathrm{Ind}_{D_s'}^D (\theta_s^{I_s'}),$$

where $D\backslash G/H$ is the set of double cosets of G modulo D and H, $D_s' = D \cap sHs^{-1}$, $I_s' = I \cap sHs^{-1}$ and $\theta_s(x) = \theta(s^{-1}xs)$ for $x \in D_s'$. (First take the invariants under I of the Mackey formula

$$\mathrm{Res}_D^G \hat\theta \simeq \bigoplus_{s \in D\backslash G/H} \mathrm{Ind}_{D_s'}^D (\theta_s)$$

[Ser1, 7.3, prop. 22], and then use the generalisation of the induction operation from [Ser1, 7.2 exer. 1].)

(c) Deduce (4.4.3) from (a) and (b).

(d) Let L/K be a Galois extension of global fields, with Galois group G. Prove the formula

$$\zeta_K(s) = \prod_\chi L(\chi, s)$$

where χ runs over all isomorphism classes of irreducible representations of G.

4.4.5 The Artin reciprocity law

Let L/K be an *abelian* extension, with Galois group G. Let $S \subset \Sigma_K$ be a finite set containing Σ_K^∞ and the set of ramified places in L. Then the Frobenius automorphism φ_v is a well-defined element of G for all $v \in \Sigma_K - S$. This gives a homomorphism

$$F_{L/K} : I_K^S \to G, \qquad (4.4.4)$$

$$v \mapsto \varphi_v.$$

This is the *Artin reciprocity map*.

Theorem 4.65 (Artin, 1927 [8]) *The pair* $(S, F_{L/K})$ *of* (4.4.4) *is admissible in the sense of Definition 4.52. Furthermore,* $F_{L/K}$ *is surjective and induces an isomorphism*

$$F_{L/K} : \mathrm{Cl}^{\mathfrak{m}}(K) \xrightarrow{\sim} G$$

for a suitable modulus $\mathfrak{m}$.

Proof See any good book on class field theory, such as Lang [Lan2] or Cassels–Fröhlich [CF]. Surjectivity follows from a generalisation of a theorem on arithmetic progressions, the *Čebotarev density theorem*. Injectivity (i.e. the existence of $\mathfrak{m}$) is much harder. $\qquad\square$

Remark 4.66 The "abstract" isomorphism of Theorem 4.65 is due to Takagi (1922, [125]); it was Artin who constructed the explicit isomorphism. We quote Tate [128, p. 315]:

How did Artin guess his reciprocity law? He was not looking for it, not trying to solve a Hilbert problem. Neither was he, as would seem so natural to us today, seeking a natural isomorphism, to make Takagi's theory more functorial. He was led to the law trying to show that a new kind of L-series which he introduced really was a generalization of the usual L-series.

This is the content of:

Corollary 4.67 *Suppose that the representation ρ of G is irreducible (i.e. of degree 1). Then*

$$L(\rho, s) = L(\rho \circ F_{L/K}, s)$$

is a Hecke L-function.

4.4.6 Local archimedean factors and the conductor

To state the functional question satisfied by $L(\rho, s)$, it is convenient to start by introducing the corresponding "completed L-function".

Start with the infinite factors when char $K = 0$. If v is a real place and $w \mid v$, then the extension L_w/K_v is of degree 1 or 2. Let G_w be its Galois group, and let φ_w be its generator (the Frobenius at infinity!): for $\varepsilon = \pm 1$, we set

$$V^\varepsilon = \{v \in V \mid \varphi_w v = \varepsilon v\}$$

(so $V^+ = V$, $V^- = 0$ if $G_w = 1$).

Definition 4.68 For $v \in \Sigma_K^\infty$,

$$\Gamma_v(\rho, s) = \begin{cases} \Gamma_{\mathbf{R}}(s)^{\dim V^+}\Gamma_{\mathbf{R}}(s + 1)^{\dim V^-} & \text{if } v \text{ is real,} \\ \Gamma_{\mathbf{C}}(s)^{\dim V} & \text{if } v \text{ is complex.} \end{cases}$$

The definition of the *Artin conductor* $\mathfrak{f}(\rho)$ [10] is much more delicate: I refer for example to Serre [Ser2, ch. VI] where it is given locally in terms of higher ramification groups (this definition depends on the *Hasse–Arf theorem*).

We now set

$$\Lambda(\rho, s) = \left(|d_K|^{\dim V} N(\mathfrak{f}(\rho)) \right)^{s/2} \prod_{v \in \Sigma_K^\infty} \Gamma_v(\rho, s) \cdot L(\rho, s).$$

4.4.7 Analytic extension and the functional equation

Theorem 4.69 (Artin–Brauer, (1947) [17]) *For every Galois representation ρ, the Artin L-function $L(\rho, s)$ extends analytically to the whole complex plane, and admits a functional equation of the form*

$$\Lambda(\rho^\vee, 1 - s) = W(\rho)\Lambda(\rho, s),$$

where $\rho^\vee$ is the dual representation of ρ and $W(\rho)$ is a complex number of modulus 1.

Proof This relies on *Brauer's theorem* ([17], [Ser1, ch. 10]): every linear representation of G is a **Z**-linear combination of induced representations of degree a representations. A version of this theorem had previously been proved by Artin, with **Z** replaced by **Q**. This proved Theorem 4.69 after raising to an integer power.

For the proof, one has to check that completed L-functions still have the induction property 4.63 of Proposition 4.63, which is done factor by factor. The proof thus reduces via Corollary 4.67 to the case of a Hecke L-function[2], and one applies Theorem 4.60.   □

Exercise 4.70 By using the formal properties of the Artin conductor, show that $\Lambda(\rho, s)$ has the same formal properties as those of Proposition 4.63.

4.4.8 Artin's conjecture

Theorem 4.69 gives us meromorphy of $L(\rho, s)$, but not

Conjecture 4.71 (Artin) *If ρ is an irreducible representation different from 1, then $L(\rho, s)$ is holomorphic in the whole complex plane.*

This conjecture remains open, except in the cases where we may take positive coefficients in Brauer's theorem and in the "automorphic" case

---

[2] In fact, one checks that Corollary 4.67 extends to complete L-functions: the point is that in the abelian case the Artin conductor coincides with the Hecke conductor of Definition 4.54.

of dimension 2 established by Langlands and his successors.[3] However, we have:

Theorem 4.72 (Weil) *Artin's conjecture holds over function fields for any nontrivial irreducible representation ρ of the Galois group G of a* regular *extension L/K, and the Riemann hypothesis is satisfied.*

The proof is in [Wei2, § V]. In fact, $Z(\rho, t)$ is a polynomial $P(t) \in \mathbf{Z}[t]$ whose inverse roots have complex absolute value $\sqrt{q}$, if q is the number of elements of the field of constants of K. The idea of the proof is to embed the group algebra $\mathbf{Z}[G]$ into the ring of correspondences $\mathrm{Corr}_{\equiv}(C, C)$ of the smooth projective model for K, and to use the positivity theorem 2.37. We will give a motivic proof of Theorem 4.72 in § 6.15.

Remark 4.73 Theorem 4.72 is false if we remove the hypothesis of "regularity" (that is to say, if K and L have different fields of constants): take for example $K = \mathbf{F}_q(x)$, $L = \mathbf{F}_{q^2}(x)$ and for ρ the character of $G = \mathrm{Gal}(L/K) = \mathbf{Z}/2$ taking value -1. Then

$$L(\rho, s) = \frac{\zeta(\mathbf{P}^1_{\mathbf{F}_{q^2}}, s)}{\zeta(\mathbf{P}^1_{\mathbf{F}_q}, s)} = \frac{(1 - q^{-2s})^{-1}(1 - q^{2-2s})^{-1}}{(1 - q^{-s})^{-1}(1 - q^{1-s})^{-1}} = \frac{1}{(1 + q^{-s})(1 + q^{1-s})}.$$

Thus, $L(\rho, s)$ has poles on the line $\Re(s) = 1$, a fact that persists if we replace $\mathbf{P}^1$ by any nonempty open subset. This contrasts with the case of number fields, where we do not require K and L to have the same number of roots of unity. See however Exercise 6.57 for a generalisation of Theorem 4.72 to the non-regular case.

4.5 The marriage of Artin and Hecke

This was accomplished by Weil, who introduced his W-groups (now called *Weil groups*): modifications of a Galois group by means of class field theory [141]; they are locally compact. In the local nonarchimedean case, or the global characteristic > 0 case, the Weil group $W(K)$ is the inverse image of $\mathbf{Z}$ (generated by Frobenius) by the continuous homomorphism $\pi : G_K \to \hat{\mathbf{Z}}$, where G_K is the absolute Galois group of K: if K is local (resp. global), π is given by the restriction to the extension of the maximal unramified extension

---

[3] There is also the theorem of Aramata–Brauer: if L/K is finite Galois, then $\zeta(L, s)/\zeta(K, s)$ is entire. The proof of this relies on Artin's theorem rather than Brauer's theorem, cf. [Ser1, 9.4, exercise].

of K (resp. to the subfield of constants). The Weil group of $\mathbf{C}$ (resp. of $\mathbf{R}$) is $\mathbf{C}^*$ (resp. the nontrivial extension of μ_2 by $\mathbf{C}^*$ realised by $\langle \mathbf{C}^*, j \rangle$ in the group of units of the field of real quaternions). In the case of number fields, the Weil group $W(L/K)$ of a finite Galois extension with Galois group G is a certain extension of G by the idele class group C_L; some highly nontrivial work is then required to show that these extensions form a coherent system "converging" towards a Weil group $W(K)$, an extension of G_K. See also Artin–Tate [AT].

This allows Weil to define L-functions that generalise both those of Hecke and Artin, and to extend Artin's factorisation of zeta functions to all Hecke L-functions. See also Tate's comments at the end of [128].

4.6 The constant of the functional equation

This section is dedicated to the constant $W(\rho)$ for the Artin L-function of a complex Galois representation ρ, appearing in the functional equation of Theorem 4.69. Two excellent references are the articles of Martinet [82] and Tate [129], of which we summarise the main points.

In his thesis [126, 4.5], Tate gives an explicit formula for $W(\rho)$ when ρ is of degree 1[4]; by Brauer's theorem, this determines the value of $W(\rho)$ for arbitrary ρ. But Tate does even better; for ρ of degree 1 he decomposes $W(\rho)$ into a product of local factors $W(\rho_v)$ depending only on the restriction ρ_v of ρ to the local Galois group G_v at a place v. More precisely, $W(\rho_v)$ is a power of i if v is real and is a certain normalised Gauss sum if v is nonarchimedean, equal to 1 if ρ is unramified at v. See loc. cit., 2.5 for the precise formulae.

Can we extend this decomposition to the general case? The answer is yes:

Theorem 4.74 (Langlands, Deligne)[5] *Let K be a local field. Then there exists a unique family of homomorphisms*

$$W : R(G_E) \to S^1$$

where E runs through the finite extensions of K, G_E is the absolute Galois group of E, $R(G_E)$ is the group of complex continuous representations of G_E, and $S^1 \subset \mathbf{C}^*$ is the unit circle, with the following properties:

---

[4] More accurately, he gives such a formula for the Hecke L-function of the character corresponding to ρ via class field theory.

[5] Langlands and Deligne prefer to express this result in terms of *epsilon factors*.

(i) *If* deg $\rho = 1$, $W(\rho)$ *is the local factor determined by Tate.*
(ii) *If* E'/E *is a finite extension and* $\rho \in R(G_{E'})$ *with* deg $\rho = 0$, *then*
$W(\text{Ind}_{G_{E'}}^{G_E} \rho) = W(\rho)$.

We have

$$W(\rho)W(\rho^*) = \det{}_\rho(-1) \qquad (4.6.1)$$

where ρ^* *is the dual representation of* ρ, *and* $\det_\rho$ *is its determinant (a representation of degree 1).*

This statement extends without difficulty to continuous representations of the Weil group (see the previous section). Uniqueness is an almost immediate consequence of Brauer's theorem. Existence is a quite different story: Langlands gave an unpublished proof of this of around 400 pages [77] before Deligne found a much shorter, global proof (see [25] and [129]).

Here are several remarkable properties of the constant $W(\rho)$.

Theorem 4.75

(1) (*Dwork, Deligne, [129, cor. 4]*) *If* ρ *is irreducible and wildly ramified,* $W(\rho)$ *is a root of unity. (This also holds in the case that* ρ *is unramified, by reduction to the abelian case.)*
(2) (*Deligne [28]*) *If* ρ *is an* orthogonal *virtual representation of* G_K *of degree* 0 *and determinant* 1, *we have*

$$W(\rho) = (-1)^{w_2(\rho)},$$

where $w_2(\rho) \in H^2(K, \mathbf{Z}/2) \simeq \mathbf{Z}/2$ *is the* second Stiefel–Whitney class *of* ρ.

Note that by virtue of (4.6.1), we also have $W(\rho) = \pm 1$ if ρ is *symplectic*; this sign is intimately linked to the Galois module structure of the ring of integers of L, where ρ factors through $\text{Gal}(L/K)$ (Fröhlich, Taylor..., see for example [22]).

Let us also point out the works of Deligne–Henniart and Henniart on the variation on $W(\rho)$ when one twists ρ by a (not too ramified) character [31, 52]. Finally, as $W(\rho)$ belongs to the maximal abelian extension A of $\mathbf{Q}$, the Galois group $\Gamma = \text{Gal}(A/\mathbf{Q})$ acts on $W(\rho)$, and its action can be compared to that of Γ on the character ρ (Fröhlich [82, cor. 5.2]); this allows one to bound the cyclotomic field containing $W(\rho)$ from above [61].

5
L-functions from geometry

5.1 "Hasse–Weil" zeta functions

5.1.1 An example of Weil

In the article [143], Weil proves a conjecture of Hasse: over a field k, let C be the algebraic plane curve given by

$$Y^e = \gamma X^f + \delta$$

where $2 \le e \le f, \gamma, \delta \in k^*$, and $ef \in k^*$. If k is a number field, we can reduce C modulo $\mathfrak{p}$ for almost all (i.e. all but finitely many) $\mathfrak{p} \subset O_K$ while preserving our hypotheses, and we call this reduction $C_\mathfrak{p}$. Write S for the set of "bad" $\mathfrak{p}$.

Definition 5.1 (Hasse)

$$\mathcal{Z}(C, s) = \prod_{\mathfrak{p} \notin S} \zeta(C_\mathfrak{p}, s).$$

Theorem 5.2 (Weil [143, p. 495]) $\mathcal{Z}(C, s)$ *admits a meromorphic extension to* **C** *and satisfies a functional equation that expresses* $\mathcal{Z}(C, 2-s)\mathcal{Z}(C, s)^{-1}$ *as a finite product of "elementary factors" (including, of course, gamma functions) which could easily be written explicitly.*

We quote Weil (scientific works, comments on [143]):

Shortly before the war, if my memory serves me well, G. de Rham told me that one of his students from Geneva, Pierre Humbert, had gone to Göttingen with the intention of working under the direction of Hasse, and that Hasse had proposed him a problem on which de Rham wanted my opinion. Given an elliptic curve C over the field of rationals, the problem had to do primarily with studying the infinite product of the zeta functions of the curves C_p obtained by reducing C modulo p for all p for which C_p is of genus 1;

more precisely, he had to determine whether this product has an analytic continuation and a functional equation (. . .) I admit that I thought Hasse had been too optimistic. (. . .)

In Brazil, while searching for applications of the Riemann hypothesis, I did some calculations with zeta functions of hyperelliptic curves over a finite field (. . .) As these notes show, it follows immediately that Hasse's infinite product for the curve $Y^2 = X^4 + 1$ is a Hecke L-function relative to the field $\mathbf{Q}(i)$. Hasse's conjecture started to take shape in front of my eyes (. . .)

(. . .) (Chevalley) did not believe that Hecke's L-functions would have a role to play in number theory. I held the opposite view (. . .) and quoted as an example Hasse's product for the curve $Y^2 = X^4 + 1$, maybe also for $Y^2 = X^3 - 1$, adding that this must be part of a much more general phenomenon. Here I was getting quite far ahead of myself; what finally gave me confidence in this idea was, once more, the analogy between number fields and function fields; I observed (. . .) that Hasse's conjecture for curves over number fields corresponded exactly to my conjectures of 1949 for surfaces over a finite field, and I no longer had any doubt about them.[1]

5.1.2 Idea of the proof

The calculations of [143] express the functions $\zeta(C_{\mathfrak{p}}, s)$ in terms of Jacobi sums: more precisely, $Z(C_{\mathfrak{p}}, t)$ is of the form

$$Z(C_{\mathfrak{p}}, t) = \prod_{a,b} L_{a,b}(t)$$

where (a, b) runs through the quotient set of $\mathbf{Z}/f \times \mathbf{Z}/e$ by the diagonal action of $(\mathbf{Z}/m)^*$, m being the least common multiple of e and f. Then, aside from the trivial cases, we have

---

[1] *Peu avant la guerre, si mes souvenirs sont exacts, G. de Rham me raconta qu'un de ses étudiants de Genève, Pierre Humbert, était allé à Göttingen avec l'intention de travailler sous la direction de Hasse, et que celui-ci lui avait proposé un problème sur lequel de Rham désirait mon avis. Une courbe elliptique C étant donnée sur le corps des rationnels, il s'agissait principalement, il me semble, d'étudier le produit infini des fonctions zêta des courbes C_p obtenues en réduisant C modulo p pour tout nombre premier p pour lequel C_p est de genre 1; plus précisément, il fallait rechercher si ce produit possède un prolongement analytique et une équation fonctionnelle. (. . .) J'avoue avoir pensé que Hasse avait été par trop optimiste. (. . .)*
Au Brésil, cherchant des applications de l'hypothèse de Riemann, je fis des calculs sur les fonctions zêta des courbes hyperelliptiques sur un corps fini (. . .) Comme l'indiquent ces notes, il s'ensuit immédiatement que le produit infini de Hasse pour la courbe $Y^2 = X^4 + 1$ est une fonction L de Hecke relative au corps $\mathbf{Q}(i)$. La conjecture de Hasse commençait à prendre forme à mes yeux. (. . .)
(. . .) (Chevalley) ne croyait pas que les fonctions L de Hecke eussent un rôle à jouer en théorie des nombres. Je soutenais le contraire (. . .) et citai en exemple le produit de Hasse pour la courbe $Y^2 = X^4 + 1$, peut-être aussi pour $Y^2 = X^3 - 1$, en ajoutant qu'il devait s'agir là d'un phénomène bien plus général. Là je m'avançais beaucoup; ce qui finalement me donna confiance dans cette idée, ce fut une fois de plus l'analogie entre corps de nombres et corps de fonctions; j'observai (. . .) que la conjecture de Hasse pour les courbes sur les corps de nombres correspond exactement à mes conjectures de 1949 pour les surfaces sur un corps fini, au sujet desquelles il ne me restait plus aucun doute.

$$L_{a,b}(t) = 1 + \lambda t^d$$

where d is the degree of a certain cyclotomic extension κ' of $\kappa(\mathfrak{p})$ and λ is essentially a *Jacobi sum*

$$j_{\mathfrak{p}}^{a,b} = \sum_{x+y=-1} \chi(x)^{a_0} \chi(y)^{b_0}.$$

This sum is calculated in κ', and χ is a multiplicative character of κ', i.e. a homomorphism $(\kappa')^* \to \mathbf{C}^*$. The key point (still simplifying somewhat) is that the rule $\mathfrak{p} \mapsto j_{\mathfrak{p}}^{a,b}$ *defines a Hecke character* $\chi^{a,b}$. The function $\mathcal{Z}(C,s)$ is then essentially the product of the L-functions of the $\chi^{a,b}$.

5.1.3 Weil's intuition

Instead of a number field, let us choose as a base field a global field k of characteristic p, say $k = \mathbf{F}_q(U)$ where U is a smooth curve. Let C be a smooth curve over k. Up to replacing U by a nonempty open set, the morphism $C \to$ Spec k extends to a smooth $\mathbf{F}_q$-morphism of finite type

$$f : C_U \twoheadrightarrow U.$$

The zeta function of C_U is then (cf. proposition 2.3 (2)):

$$\zeta(C_U, s) = \prod_{u \in U_{(0)}} \zeta(C_u, s)$$

where C_u if the fibre of f in u. On the right, we recognise a zeta function "à la Hasse", and on the left we see the zeta function of the (smooth but not projective) surface C_U.

(This typical reasoning up to a finite number of factors, analogous to the one indicated in Section 4.1.1, does not quite reduce Hasse's conjecture to Weil's conjecture for surfaces, which was only formulated in the smooth projective case: Shreeram Abhyankar did not prove resolution of singularities for surfaces over a perfect field of characteristic p until 1957 [1].)

5.1.4 General definition of Hasse–Weil L-functions: vague form

Let k be a global field, and let C be a smooth projective curve over k. Let U be a regular model of k of finite type over $\mathbf{Z}$: if k is a number field, U is of the form Spec O_S where O_S is a localisation of the ring of integers of k; if $k = \mathbf{F}_q(C)$ for a smooth projective curve C, U is a nonempty open of C. If

U is sufficiently small, then the morphism $C \to \operatorname{Spec} k$ extends to a smooth projective morphism

$$f : C_U \to U$$

as above. We can therefore, as a first approximation, define:

$$L(C, s) = \zeta(C_U, s) = \prod_{u \in U_{(0)}} \zeta(C_u, s).$$

Suppose first that $k = \mathbf{F}_q(C)$. Using Abhyankar's theorem [1], we can complete C_U into a smooth projective surface $S/\mathbf{F}_q$; then $\zeta(C_U, s) = \zeta(S, s)\zeta(S - C_U, s)^{-1}$. The factor $\zeta(S, s)$ is a rational function in q^{-s} and satisfies a functional equation linking $\zeta(S, s)$ and $\zeta(S, 2 - s)$. The closed set $S - C_U$ is of dimension 1; completing and normalising reduces to the case of smooth projective curves, and thus we obtain an (ugly) functional equation for $L(C, s)$.

Now, suppose that k is a number field. What is the generality of Weil's argument?

The point is that the curve considered by Weil (or more precisely, its projective completion) has a Jacobian J *with complex multiplication*: the ring $\operatorname{End}^0(J)$ is a commutative $\mathbf{Q}$-algebra of rank $2g$ where g is the (geometric) genus of C. In the case considered by Weil, to which he returns in [147], we have $\operatorname{End}^0(J) = \prod_{i=1}^r K_i$ where the K_i are subfields of $\mathbf{Q}(\mu_m)$ for m sufficiently large.[2] This was then generalised by Yutaka Taniyama and Weil to the case of all abelian varieties with complex multiplication [145, p. 6], [ST]. See [SGA4$\frac{1}{2}$, Sommes trig., § 5] for a reformulation of [143, 147] in terms of l-adic sheaves.

These results apply in particular in the case of elliptic curves with complex multiplication, which goes back to Deuring [36]. The case of elliptic curves without complex multiplication but defined over $\mathbf{Q}$ is much more recent: here, Hecke characters are insufficient, and (as for Artin's conjecture 4.71) one must use *modular forms* to prove the existence of an analytic continuation and the functional equation. This is the proof of the Shimura–Taniyama–Weil conjecture, stemming from the work of Andrew Wiles, Richard Taylor, Christophe Breuil, Brian Conrad, Fred Diamond (see, among others, [148, 131, 18]). These results have been generalised by Patrikis and Taylor [90], still by automorphic methods.

---

[2] Weil notes in [147] that a special case of his computations in [143] dates back to Eisenstein in 1850! [39, pp. 192–193].

5.2 Good reduction

We will see in § 5.6 how Serre proposed a precise definition of local factors of the zeta function of a smooth projective variety X defined over a global field. Serre was more demanding: he proposed a formula not just for these local factors, but also for their factorisation in terms of the cohomology of X, in the style of § 3.3. At the time of [113], this formula depended on several conjectures; among these, the Weil conjectures have since been proved, which allows for an unconditional definition of their local factors and the factorisation at least at places of good reduction. The goal of this paragraph is to remark that if we don't require such a cohomological factorisation, then the definition of these factors is *elementary*.

We start by defining the notion of good reduction.

Definition 5.3 Let K be a field equipped with a discrete valuation v of rank 1, with valuation ring $\mathcal{O}$ and residue field k. A smooth projective K-variety X has *good reduction* (relative to v) if there exists a *smooth* projective $\mathcal{O}$-scheme $\mathcal{X}$, with generic fibre $\mathcal{X}_\eta = \mathcal{X} \otimes_{\mathcal{O}} K$ isomorphic to X. We say that $\mathcal{X}$ is a *smooth model* of X, and we call $\mathcal{X} \otimes_{\mathcal{O}} k$ the *special fibre* associated to $\mathcal{X}$: this is a smooth projective k-variety.

In certain cases (K3 surfaces), one weakens the condition that $\mathcal{X}$ be a scheme by asking that it be an *algebraic space* – but the special fibre is always supposed to be a scheme.

Proposition 5.4 *Suppose that K in Definition 5.3 is a global field. Then X has good reduction at all but finitely many places of K.*

Proof We fix an embedding $X \subset \mathbf{P}^N_K$. Let $C = \operatorname{Spec} O_K$ if K is a number field, and let C be the smooth projective model of K if K is a function field. Write $\mathcal{X}$ for the closure of X in $\mathbf{P}^N_C$: this is a projective C-scheme, with generic fibre X. The morphism $p : \mathcal{X} \to C$ is smooth outside a closed set $Z \subset \mathcal{X}$; as p is proper, $p(Z)$ is closed in C and *different from* C since p is smooth at the generic point of C; so $p(Z)$ is finite. Therefore, p is smooth over the nonempty open subset $U = C - p(Z)$. $\qquad\square$

Remarks 5.5

(1) If K is a global field and v is a finite place of K, it is not true that every smooth projective K-variety has good reduction at v! The simplest

counterexample is that of an elliptic curve whose j-invariant satisfies $v(j) < 0$. (One may also take a conic without rational points, but this example is less relevant, cf. Remark 6.24 (2).)

(2) If X has a smooth model at v, then there are in general many others; given two such models $\mathcal{X}$, $\mathcal{X}'$, their special fibres Y, Y' are not in general isomorphic. Nonetheless:

Proposition 5.6 *In the situation of Remark 5.5 (2), we have equality $\zeta(Y, s) = \zeta(Y', s)$.*

Proof by anticipation. By Theorem 6.25, the fibres Y and Y' have isomorphic Chow motives, and by Lemma 6.49, the zeta function of a smooth projective k-variety only depends on its Chow motive. □

Theorem 5.7 *Let K be a global field, and let X be a smooth projective K-variety; we write $\Sigma_K(X)$ for the set of finite places of K where X has good reduction. For every place $v \in \Sigma_K(X)$, write $\zeta_v(X, s)$ for the function $\zeta(Y, s)$, where Y is the special fibre of an arbitrary smooth model of X at v (Proposition 5.6). Then the product*

$$\prod_{v \in \Sigma_K(X)} \zeta_v(X, s)$$

converges absolutely for $\Re(s) > \dim X + 1$.

Proof We reuse the notation from the proof of Proposition 5.4: the set $U_{(0)}$ of closed points of the open U is contained in $\Sigma_K(X)$, and the complement $\Sigma_K(X) - U_{(0)}$ is finite. As $\zeta_v(X, s)$ converges absolutely for $\Re(s) > \dim X$ (Theorem 2.5), it suffices to prove convergence with $\Sigma_K(X)$ replaced by $U_{(0)}$. But we have

$$\prod_{v \in U_{(0)}} \zeta_v(X, s) = \zeta(p^{-1}(U), s)$$

(Proposition 2.3), and the second term converges absolutely for $\Re(s) > \dim p^{-1}(U) = \dim X + 1$ by reapplying Theorem 2.5. □

Theorem 5.7 provides a definition for the "unramified part" of the Hasse–Weil zeta function of X.

5.3 *L*-functions of *l*-adic sheaves

References: Grothendieck [43], [SGA5, exp. XV], and Illusie [55] for very enlightening historical comments (which go far beyond the scope of *L*-functions).

5.3.1 *l*-adic sheaves

We quote Serre [CorrGS, end of letter of 26-10-61]:

(About affine varieties: can you decompose (filter, or whatever) their cohomology in such a way as to highlight the pieces that arise from the complete variety? I can't manage to express myself well (*), but you will understand what I want if I put it in this form: what conjectures should one make for the zeta function of a nonsingular affine variety? I find it outrageous that one should have to embed it (if at all possible) into a projective nonsingular variety, not at all unique, and gross; on the other hand, I cannot formulate anything. Do you have another homology than the usual at hand (for example "with closed support" or God knows what) which could be of use?)
(*) I lacked the language of motives. (note added in 2001).[3]

and then Grothendieck [43]:

The goal of this exposition is to prove the rationality result for all L-functions considered in § 1, and even for a much more general type of L-functions associated to l-adic *sheaves on X. In fact, we will give an explicit Lefschetz–Weil style formula for these functions. The essential tools come in two forms:*
(a) The formalism of "cohomology with compact support" *(...)*
(b) A generalised Lefschetz formula, *due to* J. L. Verdier (...).[4]

---

[3] *(À propos des variétés affines: peux-tu décomposer (filtrer, ou n'importe quoi) leur cohomologie de façon à mettre en évidence les morceaux qui proviennent de la variété complète? Je n'arrive pas à m'exprimer (*), mais tu comprendras ce que je veux si je le mets sous la forme: quelles conjectures doit-on faire pour la fonction zêta d'une variété affine non singulière? Il me paraît scandaleux qu'il soit nécessaire de la plonger (si tant est que ce soit possible!) dans une variété projective non singulière, pas du tout unique, et dégueulasse; d'autre part, je n'arrive à rien formuler. As-tu en main une autre homologie que l'usuelle (par exemple «à supports fermés» ou Dieu sait quoi) qui pourrait servir à quelque chose?)*
(*) Il me manquait le langage des motifs. (note ajoutée en 2001).

[4] *Le but de cet exposé est de prouver le résultat de rationalité pour toutes les fonctions L envisagées au § 1, et même pour un type de fonctions L beaucoup plus général, associé à des* faisceaux *l*-adiques *sur X. Nous donnerons en effet une formule explicite du style Lefschetz-Weil de ces fonctions. Les outils essentiels sont de deux sortes:*
(a) Le formalisme de la «cohomologie à supports compacts»/ *(...)*
(b) Une formule de Lefschetz généralisée, *due à* J. L. Verdier (...).

Here are the *l*-adic sheaves to which Grothendieck alluded: we suppose in what follows that the reader is familiar with the basics of étale cohomology (see [SGA4] or [Mil]).

Definition 5.8 ([SGA5, exp. VI, def. 1.1.1 and 1.2.1]) Let X be a scheme, and let l be a prime number.

a) An *l-adic sheaf* on X is a projective system $F = (F_n)_{n\geq 0}$ of étale sheaves on X such that, for all n, F_n is annihilated by l^{n+1}, and the projection morphism $F_{n+1} \to F_n$ induces an isomorphism $F_{n+1} \otimes_{\mathbf{Z}} \mathbf{Z}/l^{n+1} \xrightarrow{\sim} F_n$.

b) The sheaf F is *constructible* if each F_n is constructible (for this, it suffices that F_0 be constructible).

c) The sheaf F is *twisted-constant* [constructible][5] if the F_n are locally constant and constructible.

When X is connected, if $a \to X$ is a geometric point, then we have an equivalence of categories [SGA5, exp. VI, prop. 1.2.5]

$$\{\text{twisted-constant } l\text{-adic sheaves}\} \leftrightarrows$$
$$\{\text{continuous representations of } \pi_1(X, a)$$
$$\text{in finite type } \mathbf{Z}_l\text{-modules}\}. \quad (5.3.1)$$

Warning 5.9 A continuous representation of a profinite group over a finite-dimensional **C**-vector space factors through a finite quotient. This is totally false if we replace the **C**-vector space by a $\mathbf{Z}_l$-module, or even by a $\mathbf{Q}_l$-vector space (example: $\mathbf{Z}_l$ acts continually and faithfully on $\mathbf{Q}_l$ by $x \mapsto \exp(lx)$).

Exercise 5.10 (Cyclotomic character) Let k be a field with separable closure k_s. If $G = \mathrm{Gal}(k_s/k)$ and m is an integer invertible in k, the action of G on the roots of unity μ_m defines a homomorphism

$$\kappa^{(m)} : G \to \mathrm{Aut}(\mu_m) \simeq (\mathbf{Z}/m)^*;$$

this is the *cyclotomic character modulo m*. By passing to the projective limit, we get the cyclotomic character

$$\kappa : G \to \varprojlim(\mathbf{Z}/m)^* = \prod_{l \neq p} \mathbf{Z}_l^*,$$

where $p = \mathrm{char}\, k$. By projecting onto $\mathbf{Z}_l^*$, we obtain its *l*-primary component κ_l.

---

[5] Later, we will also say *lisse*.

(a) Suppose $k = \mathbf{F}_q$, and let $\varphi_q \in G$ be the arithmetic Frobenius ($\varphi_q(x) = x^q$). Show that $\kappa_l(\varphi_q) = q$ for all $l \neq p$. Deduce that the image of κ_l is open in $\mathbf{Z}_l^*$.

(b) Suppose that $k = \mathbf{Q}$. Show that κ_l is surjective for all l. (Use the irreducibility of cyclotomic polynomials.)

(c) Suppose that k is of finite type over its prime subfield. Show that the image of κ_l is open.

5.3.2 Direct and inverse images...

Let $f : X \to Y$ be a morphism of schemes. I refer to [SGA4] or Milne [Mil, ch. III] for the definition of the inverse image f^* of sheaves and the higher direct images $R^q f_*$, and especially for the higher direct images with proper supports $R^q f_!$ (these last functors for when f is a finite type separated morphism of locally Noetherian schemes), whose complete construction is carried out in Exposé XVII of [SGA4] (see also [Mil, ch. VI, § 3]).

Following [SGA5, exp. VI], we may define these functors on l-adic sheaves[6], and we have:

Theorem 5.11 ([SGA5, exp. VI, Lemma 2.2.2 and 2.2.3 A)]) *If F is constructible, then $R^q f_! F$ is constructible for all $q \geq 0$ and zero for $q > 2d$, where d is the relative dimension of f.*

Theorem 5.12 ([SGA5, exp. VI, 2.2.3 B)]) *The formation of the $R^q f_! F$ commutes with base change.*

We elaborate on this last point using the language of derived categories (§ B.2.5). In general, let

$$
\begin{array}{ccc}
\mathcal{C}' & \xrightarrow{g'_*} & \mathcal{C} \\
f'_* \downarrow & & f_* \downarrow \\
\mathcal{D}' & \xrightarrow{g_*} & \mathcal{D}
\end{array}
$$

be a naturally commuting diagram of categories and functors: naturally commuting means that there exists a natural isomorphism

$$f_* g'_* \simeq g_* f'_*. \tag{5.3.2}$$

---

[6] This point is somewhat delicate, since in general the direct image of an l-adic sheaf doesn't quite satisfy the condition of Definition 5.8 a). The way to rectify this is by using the technique of Theorem B.7 with the multiplicative system of the shift morphisms $(F_n) \mapsto (F_{n+r})$, cf. [SGA5, exp. V, §§ 2.4 and 2.5].

Suppose that g_* and g'_* have left adjoints g^*, g'^*. From (5.3.2), we get a composition

$$f_* \xrightarrow{f_* * \eta'} f_* g'_* g'^* \simeq g_* f'_* g'^*$$

where η' is the unit of the adjunction (g'^*, g'_*). Then

$$g^* f_* \to g^* g_* f'_* g'^* \xrightarrow{\varepsilon * f'_* g'^*} f'_* g'^* \qquad (5.3.3)$$

where ε is the counit of the adjunction (g^*, g_*): this is the *base change morphism*. When this situation arises from a cartesian square of $\mathbf{F}_q$-schemes of finite type

$$
\begin{array}{ccc}
X' & \xrightarrow{g'} & X \\
f' \downarrow & & \downarrow f \\
Y'' & \xrightarrow{g} & Y
\end{array}
\qquad (5.3.4)
$$

with $\mathcal{C} = D^b_c(X, \mathbf{Z}_l)$, etc. and f_* meaning Rf_*, etc. (see §§ 5.4.2 and 5.4.3 below), the proper base change theorem implies that (5.3.3) is an isomorphism when f is proper, for arbitrary g.

If $f : X \to Y$ is now an arbitrary morphism of $\mathbf{F}_q$-schemes of finite type, it can be factored as

$$X \xrightarrow{j} \bar{X} \xrightarrow{\bar{f}} Y$$

where j is an open immersion and $\bar{f}$ is proper (thanks to Nagata's compactification theorem[7]). By definition, we have

$$Rf_! = R\bar{f}_* \circ j_!,$$

where the proper base change theorem ensures that this definition is independent of the choice of $(j, \bar{f})$. We then define the base change isomorphism

$$g^* Rf_! \xrightarrow{\sim} Rf'_! g'^* \qquad (5.3.5)$$

as the composition of (5.3.3) for $\bar{f}$ and the trivial isomorphism

$$g''^* j_! \xrightarrow{\sim} j'_! g'^*.$$

(Here, g'' is the base change of g by $\bar{f}$ and j' is the base change of j by g''.)

---

[7] If $X \subset \bar{X}_0$ is a compactification of X over $\mathbf{F}_q$, we may take for $\bar{X}$ the closure in $\bar{X}_0 \times Y$ of the graph of f in $X \times Y$.

5.3.3 The L-function of a constructible l-adic sheaf

Definition 5.13 ([SGA5, exp. XV, § 3 n° 1)]) a) Let F be an l-adic sheaf on $\operatorname{Spec} \mathbf{F}_q$ ($l \nmid q$), that is to say, an l-adic representation of $\operatorname{Gal}(\bar{\mathbf{F}}_q/\mathbf{F}_q) \simeq \hat{\mathbf{Z}}$. We define

$$L(\mathbf{F}_q, F) = \det(1 - \varphi_q^{-1} t^n \mid F \otimes \mathbf{Q}_l)^{-1} \in \mathbf{Q}_l(t) \subset \mathbf{Q}_l[[t]]$$

where φ_q is the Frobenius automorphism $x \mapsto x^q$ of $\bar{\mathbf{F}}_q$ and $q = p^n$.

b) Let X be a scheme of finite type over $\mathbf{F}_q$, and let F be a constructible l-adic sheaf on X. We define:

$$L(X, F) = \prod_{x \in X_{(0)}} L(x, F_x),$$

where F_x is the fibre of F in x (i.e. $i_x^* F$, where $i_x : x \to X$ is the canonical closed immersion).

5.3.4 Explanation of φ_q^{-1}

Let $\pi_X : X \to X$ be the Frobenius endomorphism over $\mathbf{F}_q$: it induces an isomorphism of l-adic sheaves $(\pi_X)_* F \xrightarrow{\sim} F$, whose inverse defines, after adjunction, a morphism

$$Fr_{F/X}^* : \pi_X^* F \to F$$

known as the "Frobenius correspondence". One "trivially" checks that the composition

$$H^*(X, F) \to H^*(X, \pi_X^* F) \to H^*(X, F)$$

is equal to the identity, where the first arrow is the canonical map (defined by the morphism $F \to (\pi_X)_* \pi_X^* F$) and the second arrow is $H^*(X, Fr_{F/X}^*)$. By applying this to $\bar{X} = X \times_{\mathbf{F}_q} \bar{\mathbf{F}}_q$, we deduce that the action of $\pi_X \times 1$ on $H^*(\bar{X}, Fr_{F/X}^*)$ is the *inverse* of that of $1 \times \pi_{\mathbf{F}_q}$, that is, the action of φ_q: *the arithmetic Frobenius is the inverse of the geometric Frobenius.*

For an excellent exposition of this incomprehensible lemma, see Houzel's exposé [SGA5, exp. XV, §§ 1 and 2].

5.3.5 Convergence; functoriality in F

Lemma 5.14 *The infinite product $L(X, F)$ is convergent in $\mathbf{Q}_l[[t]]$ (as a formal power series).*

Proof This comes down to saying that the set

$$\{x \in X_{(0)} \mid v_t(L(x, F_x) - 1) \le n\}$$

is finite for all $n > 0$, which follows from the definition (cf. Exercise 4.64 (a)). □

Lemma 5.15 *If* $0 \to F' \to F \to F'' \to 0$ *is an exact sequence of constructible l-adic sheaves, we have*

$$L(X, F) = L(X, F')L(X, F'').$$

Proof This is clear, by reducing to the case of finite fields. □

5.3.6 The theorem

Theorem 5.16 ([SGA5, exp. XV, §3 n° 2)]) *Let* $f : X \to Y$ *be an* $\mathbf{F}_q$-*morphism of finite type. For any l-adic sheaf F on X, we have*

$$L(X, F) = \prod_{i=0}^{2n} L(Y, R^i f_! F)^{(-1)^i}.$$

By taking $Y = \operatorname{Spec} \mathbf{F}_q$, we get:

Corollary 5.17 $L(X, F) = \prod_{i=0}^{2n} \det(1 - \varphi_q^{-1}t \mid H_c^i(\bar{X}, F) \otimes_{\mathbf{Z}_l} \mathbf{Q}_l)^{(-1)^{i+1}}$ *is a rational function (with coefficients in* $\mathbf{Q}_l$*).*

Corollary 5.18 *We have* $Z(X, t) = \prod_{i=0}^{2n} \det(1 - \varphi_q^{-1}t \mid H_c^i(\bar{X}, \mathbf{Q}_l))^{(-1)^{i+1}}$ *and* $Z(X, t) \in \mathbf{Q}_l(t) \cap \mathbf{Q}[[t]] = \mathbf{Q}(t)$ *(cf. Exercise 3.71).*

Indeed, $Z(X, t) = L(X, \mathbf{Z}_l)$.

In the following sections, we give an idea of the proof of Theorem 5.16. Since there are at least three complete expositions [SGA4$\frac{1}{2}$, SGA5, Mil], we content ourselves with a cursory overview.

5.3.7 Reduction to $X =$ an open subset of $\mathbf{P}^1$ (loc. cit.)

In four steps:

(i) If Theorem 5.16 holds for each fibre of f over a closed point of Y, then it holds for f (this follows from Theorem 5.12).

(ii) The theorem holds if f is finite. (By 1, we are reduced to an extension of finite fields $\mathbf{F}_{q'}/\mathbf{F}_q$, and then

$$L(\mathbf{F}_{q'}, F) = L(\mathbf{F}_q, \operatorname{Ind}_{G_{\mathbf{F}_{q'}}}^{G_{\mathbf{F}_q}} F)$$

follows from the same calculation as in the proof of Proposition 4.63 (iii).)

(iii) The theorem is stable under composition of morphisms (this follows from the Leray spectral sequence

$$R^p g_! R^q f_! F \Rightarrow R^{p+q} (g \circ f)_! F$$

and Lemma 5.15).[8]

(iv) Suppose $Y = \operatorname{Spec} \mathbf{F}_q$. If Z is a closed subset of X with complement U and if the theorem holds for $f_{|Z}$, then it holds for f if and only if it holds for $f_{|U}$. (This follows from Lemma 5.15 and the exact sequence of cohomology with proper support:

$$\cdots \to R^q(f_{|U})_! F \to R^q f_! F \to R^q (f_{|Z})_! F \to R^{q+1}(f_{|U})_! F \to \cdots)$$

Using point (iv) and Noetherian induction, we reduce to the case of X affine, and then by point (ii) and the Noether normalisation lemma to $X = \mathbf{A}^n$, and then by points (i) and (iii) to $n = 1$. We then see easily that Corollary 5.17 holds for $\dim X = 0$: by reusing point (iv), we may replace $\mathbf{A}^1$ by an arbitrary nonempty open U of $\mathbf{P}^1$.

5.3.8 Proof for $X = U \subset \mathbf{P}^1$

By taking logarithms, Corollary 5.17 is equivalent via (3.6.2) to the *trace formula*

$$\sum_{x \in U_{(0)}} \operatorname{Tr}(\varphi_q^{-n} \mid F_x) = \sum_{i=0}^{2} (-1)^i \operatorname{Tr}(\varphi_q^{-n} \mid H_c^i(\bar{U}, F) \otimes \mathbf{Q}_l). \qquad (5.3.6)$$

Since $\varphi_q^n = \varphi_{q^n}$, up to replacing $\mathbf{F}_q$ by $\mathbf{F}_{q^n}$ we may suppose that $n = 1$. Furthermore, up to restricting U, we may suppose that F is *twisted-constant*. If F is actually constant, we easily reduce the proof of (5.3.6) to the one-dimensional case of the trace formula (3.6.1), and this holds even when replacing $\mathbf{P}^1$ by an arbitrary smooth projective curve. We would like to deduce the general case via the equivalence of categories (5.3.1), but the covering of U defined by F is infinite in general (see 5.9). For this reason, one reduces

---

[8] The point is this: "spectral sequences preserve Euler–Poincaré characteristics"; for this type of reasoning, see for example Serre [Ser4, p. V-24].

to proving an analogue of (5.3.6) for a *locally free sheaf F of* $\mathbf{Z}/l^n$-*modules* using a very delicate argument.

The first pitfall is that the meaning of the second term is not *a priori* clear, since the $\mathbf{Z}/l^n$-modules $H^i_c(\bar{U}, F)$ are not free in general. Furthermore, we cannot hope to replace these by finite projective resolutions, since the ring $\mathbf{Z}/l^n$ has infinite homological dimension for $n > 1$. The solution is to consider the $H^i_c(\bar{U}, F)$ globally: it turns out that $H^*_c(\bar{U}, F)$ is the cohomology of a bounded complex K of free $\mathbf{Z}/l^n$-modules on which the φ_q operate (see for example [Mil, ch. VI, lemma 8.15]), which is sufficient; K is an example of a *perfect complex* (see Definition B.36 for a more general definition).

The argument of [SGA5, exp. XV, § 3 n° 3] also uses subtle lemmas from homological algebra over the rings $\mathbf{Z}/l^n$.

There are then two proofs:

- A "Lefschetz–Verdier formula" which computes the second term of (5.3.6) as a sum of local terms, which are then identified with the $\mathrm{Tr}(\varphi_q^{-n} \mid F_x)$. A short but detailed exposition can be found in [91][9].
- A reduction to the trace formula from Proposition 3.49 inspired by the work of Nielsen and Wecken [137], which we will now summarise. (References: [SGA5, exp. XII], [SGA4$\frac{1}{2}$, Rapport], [Mil, ch. VI, § 13].)

Since F is locally constant constructible, it is a $\pi_1(U, u)$-module for a geometric point $u \to U$; this action factors through the Galois group G of a finite (connected) étale covering $\pi : U' \to U$. We would like to prove the formula

$$\sum_{x \in U_{(0)}} \mathrm{Tr}(\varphi_q^{-1} \mid F_x) = \mathrm{Tr}(\varphi_q^{-1} \mid R\Gamma_c(\bar{U}, F)). \tag{5.3.7}$$

Let K (resp. K') be the function field of U (resp. of U'): the extension K'/K is Galois with Galois group G. Let $k = \mathbf{F}_q$ (resp. k') be the algebraic closure of $\mathbf{F}_p$ in K (resp. in K'): we have a diagram of Galois extensions

$$
\begin{array}{ccccc}
k' & \longrightarrow & K' & \longrightarrow & K'\bar{k} \\
{\scriptstyle g}\uparrow & & {\scriptstyle G}\uparrow & & {\scriptstyle H}\uparrow \\
k & \longrightarrow & K & \longrightarrow & K\bar{k}
\end{array}
$$

where the group H is the Galois group of the geometric étale covering $\bar{U}' = U' \times_{k'} \bar{k} \to U \times_k \bar{k} = \bar{U}$. The right part of the above diagram is the generic fibre of the (not necessarily cartesian!) square

---

[9] I thank Oussama Ouriachi for showing me this reference.

$$
\begin{array}{ccc}
U' & \longleftarrow & \bar{U}' \\
{\scriptstyle \pi}\downarrow{\scriptstyle G} & & {\scriptstyle \bar{\pi}}\downarrow{\scriptstyle H} \\
U & \xleftarrow[\Gamma]{} & \bar{U}
\end{array}
$$

where $\Gamma = \mathrm{Gal}(\bar{k}/k)$. We have a commutative diagram of exact sequences of profinite groups

$$
\begin{array}{ccccccccc}
1 & \longrightarrow & H & \longrightarrow & \tilde{G} & \xrightarrow{\ p\ } & \Gamma & \longrightarrow & 1 \\
& & \| & & \downarrow & & \downarrow & & \\
1 & \longrightarrow & H & \longrightarrow & G & \longrightarrow & g & \longrightarrow & 1
\end{array}
$$

where $\tilde{G} = \mathrm{Gal}(\bar{U}'/U)$ (the vertical arrows are surjective).

The group Γ is isomorphic to $\hat{\mathbf{Z}}$, with generator φ^{-1}, the inverse of the "arithmetic" Frobenius $\varphi := \varphi_q$. Write $H_1 = p^{-1}(\varphi^{-1}) \subset \tilde{G}$: this is an H-torsor whose elements operate on $\bar{U}'$.

Next, for every $\mathbf{F}_q$-scheme X of finite type we write π_X for the "geometric" Frobenius of $\bar{X}$, so that the action of π_X on $R\Gamma_c(\bar{X}, F)$ is equal to that of φ^{-1} (cf. § 4.4.2). We then have actions of H_1 and $\pi_{U'}$ on $\bar{U}'$: the first is over U and the second over $\bar{k}$.

Lemma 5.19 *Let $E = \{\sigma \in \mathrm{Aut}_{\bar{k}}(\bar{U}') \mid \bar{\pi}\sigma = \pi_U\bar{\pi}\}$. Then*

$$
H_1 = E = \pi_{U'}H. \qquad \qquad \square
$$

The group H acts on H_1 by conjugation, so also on H by transport of structure: let S be the set of orbits of this action. For $h \in H$, write Z_h for its stabiliser: up to conjugation, it only depends on the image of h in S. The fundamental formula is then ([SGA4$\frac{1}{2}$, Rapport, (5.12.1)], [Mil, ch. VI, lemma 13.15]):

$$
\mathrm{Tr}(\varphi_q^{-1} \mid R\Gamma_c(\bar{U}, F)) = \mathrm{Tr}(\pi_U \mid R\Gamma_c(\bar{U}, F))
$$

$$
= \sum_{h \in S} \frac{1}{|Z_h|} \mathrm{Tr}(\pi_{U'}h^{-1} \mid H_c^*(\bar{U}', \mathbf{Q}_l)) \cdot \mathrm{Tr}(h \mid \pi^*F) \quad (5.3.8)
$$

(see § 5.3.4 for the first equality), where the l-adic numbers

$$
\frac{1}{|Z_h|} \mathrm{Tr}(\pi_{U'}h^{-1} \mid H_c^*(\bar{U}', \mathbf{Q}_l))
$$

belong to $\mathbf{Z}_l$ and π^*F is considered as a *constant* sheaf on U', equipped with a G-action (and therefore an H-action).

Let C' be the smooth projective completion of U': the action of G extends canonically to C'; note that the fixed points of $\pi_{C'}h^{-1}$ on $\bar{C}'$ are all of multiplicity 1. By applying the trace formula from Proposition 3.49, a short computation gives Theorem 5.16 (again using point (iv) of § 5.3.7).

To prove formula (5.3.8), we need the notion of *noncommutative trace* (Stallings [121], Hattori). Let R be a unitary ring, not necessarily commutative. We write

$$H_0(R) = R/[R, R]$$

for the quotient of the additive group of R by the subgroup generated by commutators $ab - ba$: this is the 0th Hochschild homology group of R. Given a matrix $M = (m_{ij}) \in M_n(R)$, we define its trace to be

$$\mathrm{Tr}_R(M) = \sum m_{ii} \in H_0(R).$$

This definition extends to endomorphisms of projective left R-modules of finite type. One shows that, if C is a perfect complex on R and f is an endomorphism of C in $D(R)$, then the trace of f is well-defined [SGA4$\frac{1}{2}$, Rapport, 4.3]. Then (5.3.8) follows from formal properties of the noncommutative trace, the most important one being:

Lemma 5.20 ([SGA4$\frac{1}{2}$, Rapport, prop. 5.6]) *Let H be a finite group, and let P be a projective $\Lambda[H]$-module, where Λ is a ring (for example $\Lambda = \mathbf{Z}/l^n$). Let u be an endomorphism of P. Then*

$$\mathrm{Tr}_\Lambda(u) = |H|\,\mathrm{Tr}_{\Lambda[H]}(u).$$

5.3.9 Generalisations: perfect complexes and Weil sheaves

First of all, we can define the L-function of a perfect complex C of l-adic sheaves on X:

$$L(X, C) = \prod_{i \in \mathbf{Z}} L(X, H^i(C))^{(-1)^i}.$$

This definition depends only on the class of C in $D_c^b(X, \mathbf{Q}_l)$ (see § 5.4.3). Theorem 5.16 can then be more elegantly reformulated as:

$$L(X, C) = L(Y, Rf_!C). \tag{5.3.9}$$

Another generalisation is to *Weil sheaves*. For X of finite type over $\mathbf{F}_p$, with $\bar{X} = X \times_{\mathbf{F}_p} \bar{\mathbf{F}}_p$, a Weil sheaf on X is an étale sheaf F on $\bar{X}$ equipped with a Frobenius action, i.e. a morphism $\pi_F : \pi_X^* F \to F$. An étale sheaf on X is a particular case of a Weil sheaf. We use the adjectives constructible, l-adic, etc. if these apply to F on $\bar{X}$.

We can then define the L-function of a Weil sheaf, or even a perfect complex of such sheaves, and then state the same theorem, with the same proof [25, § 10].

The idea of Weil sheaves was taken up by Lichtenbaum to define his *Weil-étale topology*, and the corresponding cohomology [79].

5.4 The functional equation in characteristic p

In a 1964 letter to Serre, Grothendieck sketched the proof of functional equations for the L-functions we have just considered. The goal of this section is to discuss the details. For this, we need several preliminaries.

5.4.1 The six operations formalism

To express a functional equation satisfied by the L-function of an l-adic sheaf, we need the six operations formalism

$$f^*, f_*, f_!, f^!, \otimes, \underline{\mathrm{Hom}}$$

invented by Grothendieck. Let us first deal with

Definition 5.21 Let S be a scheme, and let $\mathrm{Sch}(S)$ be the category of S-schemes of finite type (the morphisms are morphisms of S-schemes). A *four operations formalism* on S is the data:

(1) For every $X \in \mathrm{Sch}(S)$, a category $T(X)$.
(2) For every morphism $f : X \to Y$ in $\mathrm{Sch}(S)$, 4 functors

$$f^*, f^! : T(Y) \to T(X), \quad f_*, f_! : T(X) \to T(Y)$$

equipped with natural coherent isomorphisms $(g \circ f)^* \simeq f^* \circ g^*$, etc., with the following properties

(a) f_* is right adjoint to f^* and $f_!$ is left adjoint to $f^!$.

(b) **Weak gluing**: Let $Z \overset{i}{\to} X \overset{j}{\leftarrow} U$ be a decomposition of X into a closed set (Z) and its open complement (U). Then the pair $(i^*; j^*)$ is conservative: if $\varphi : x \to y$ is an arrow of $T(X)$ such that $i^*\varphi$ and $j^*\varphi$ are isomorphisms, then φ is an isomorphism.

(c) We have a morphism of functors $f_! \to f_*$, which is an isomorphism if f is proper.

(d) If f is smooth, there exists an auto-equivalence $Th(f)$ of $T(Y)$ such that $f^! \simeq Th(f) \circ f^*$, with the "obvious" compatibilities for the composition of smooth morphisms. If f is étale, then $Th(f) = 1$.

For the last two operations, suppose that $T(X)$ is equipped with the structure of a closed monoidal category for all X, in general unitary and symmetric[10] (§ A.2, Definition A.20). We would like the following properties:

M1 The f^* are monoidal functors.

M2 We have canonical "associative and unitary" pairings

$$f^*a \otimes f^!b \to f^!(a \otimes b)$$

which are isomorphisms for smooth f; in this case, the auto-equivalence $Th(f)$ of Definition 5.21 (2) ((2)(d)) is obtained by setting $b = \mathbf{1}$ above.

M3 The morphism of functors

$$f_!(f^*a \otimes c) \to a \otimes f_!c$$

induced from M2 by adjunction is an isomorphism (the projection formula).

We then formally obtain an isomorphism of functors

M4 $\underline{\mathrm{Hom}}(f_!a, b) \xrightarrow{\sim} f_* \underline{\mathrm{Hom}}(a, f^!b)$.

Let $\mathcal{M}$ be a closed unital symmetric monoidal category, and let K be an object of $\mathcal{M}$. For all $X \in \mathcal{M}$, the evaluation morphism (the counit of the adjunction $\otimes - \underline{\mathrm{Hom}}$)

$$X \otimes \underline{\mathrm{Hom}}(X, K) \to K$$

gives, by adjunction, a *biduality arrow*

$$X \to \underline{\mathrm{Hom}}(\underline{\mathrm{Hom}}(X, K), K). \qquad (5.4.1)$$

Definition 5.22 We say that K is a *dualising object* if (5.4.1) is an isomorphism for all $X \in \mathcal{M}$.

A dualising object is not uniquely determined: if K is dualising, then so is $L \otimes K$ for any invertible object L of $\mathcal{M}$ (i.e. such that there exists L^{-1} and an isomorphism $\mathbf{1} \xrightarrow{\sim} L \otimes L^{-1}$). The converse holds too: if K' is dualising, $L = \underline{\mathrm{Hom}}(K, K')$ is invertible with inverse $\underline{\mathrm{Hom}}(K', K)$, and $K \otimes L \xrightarrow{\sim} K'$.

---

[10] Though Ayoub avoids this last hypothesis in [Ayo, ch. 2].

We would then like the following properties:

M5 If $K \in T(Y)$ is dualising, then $f^! K \in T(X)$ is dualising for all $f :$ $X \to Y$.

M6 If X is regular, $\mathbf{1} \in T(X)$ is dualising.

Suppose that S is regular and satisfies **M6**. Let $f : X \to S$ be a smooth morphism: by **M2**, $f^! \mathbf{1}$ is invertible and the conditions **M5** and **M6** are equivalent for X.

Still supposing that S is regular: we see

$$K_X = f_X^! \mathbf{1}$$

for all $X \in \mathrm{Sch}(S)$, where $f_X : X \to S$ is the structure morphism, and

$$D_X(a) = \underline{\mathrm{Hom}}(a, K_X)$$

for $a \in T(X)$. Supposing **M1**,..., **M6** hold, D_X is a biduality for all $X \in$ $\mathrm{Sch}(S)$ and one formally shows the following identities for an S-morphism $f : X \to Y$:

$$D_X(f^* y) \simeq f^! D_Y(y), \quad D_X(f^! y) \simeq f^* D_Y(y), \quad (5.4.2)$$

$$D_Y(f_* x) \simeq f_! D_X(x), \quad D_Y(f_! x) \simeq f_* D_X(x). \quad (5.4.3)$$

In practice, we ask that the $T(X)$ be triangulated categories and that the functors f^*, etc. be triangulated functors.

Remark 5.23 I have tried to give a list of the properties that hold in all the cases I know, but I don't attempt here to give a minimal system of axioms, or a maximal collection of properties, or (except in easy cases) an idea of the possible proofs of these consequences. This would not only be outside the scope of these notes, it even seems impossible at present. To my knowledge there are four formalisms of six operations in algebraic geometry:

(1) Grothendieck's coherent duality, explained by Robin Hartshorne in [Har1] and taken up by Brian Conrad [Con], then Amnon Neeman [87]. (But $j_!$ does not exist for an open immersion j...)

(2) Duality in étale or l-adic cohomology, which is the object of [SGA4] and [SGA5] via [SGA4$\frac{1}{2}$] (see below).

(3) Duality in the motivic world à la Voevodsky, constructed by Joseph Ayoub in [Ayo].

(4) Duality for holonomic D-modules and for mixed Hodge modules (for example [102]).

In these four cases, the principles of the starting axioms and the proofs resemble each other without quite overlapping. One starts from "easy" data, that is to say, f^* and f_* in the cases (1) and (2), and one constructs the others based on nontrivial theorems of base change and finitude. But the strategies are different.

In the étale case one has base change for all proper morphisms and by all smooth morphisms; in the coherent case, one has a base change theorem for all flat morphisms, whereas for proper morphisms one only has a semicontinuity theorem.

Case (3) is still different: one uses additionally the (easy) information of a left adjoint of f^* when f is smooth, and then properties of A^1-homotopy invariance which are forced axiomatically (while they are deduced from the theorem in case (2) and are false in case (1)). Contrary to the étale case, base change by a smooth morphism is easy to prove, and base change for a proper morphism follows by using the axiom of weak gluing (Definition 5.21(2)(b)), which is the least formal property.

I say nothing of case (4), by lack of competence.

It seems (to the author) that clarifying the links between these constructions and revealing a possible common skeleton could be a useful task to undertake.

5.4.2 The six operations in étale cohomology

For any scheme X, any prime number l invertible in X and any integer $n \geq 1$, we write

$$D_c^b(X, \mathbf{Z}/l^n) = D_c^b(X)$$

for the full subcategory of $D(X_{\text{ét}}, \mathbf{Z}/l^n)$ (the derived category of étale sheaves of $\mathbf{Z}/l^n$-modules) formed of those bounded complexes C such that $H^i(C)$ is constructible for all $i \in \mathbf{Z}$. This is a thick subcategory of $D(X_{\text{ét}}, \mathbf{Z}/l^n)$ in the sense of Definition B.20 b).

Let $f : X \to Y$ be a morphism of schemes. It is clear that

$$f^* D_c^b(Y) \subset D_c^b(X);$$

the theorem of finitude for cohomology with proper supports assures the existence of a triangulated functor

$$Rf_! : D_c^b(X) \to D_c^b(Y)$$

when f is compactifiable[11]. What about the other operations f_*, $f^!$, <u>Hom</u>, $\otimes$?

---

[11] Or, more generally, separated of finite type and such that X and Y are quasicompact and separated: [SGA4, exp. XVII, § 7].

Theorem 5.24 (Deligne, [SGA4$\frac{1}{2}$, Th. finitude, th. 1.1]) *Let S be a regular scheme of dimension $\leq$ 1, and let* Sch(S) *be the category of S-schemes of finite type. Let l be a prime number invertible on S. Let $f : X \to Y$ be a morphism of* Sch(S), *and let F be a constructible étale l^n-torsion sheaf on X. Then $R^q f_* F$ is constructible for all $q \geq 0$ and zero for $q \gg 0$. Consequently:*

$$Rf_* D_c^b(X) \subset D_c^b(Y).$$

Corollary 5.25 (ibid., cor. 1.5, cor. 1.6) *Under the same hypotheses,*

$$Rf^! D_c^b(Y) \subset D_c^b(X); \quad \text{R\underline{Hom}}(D_c^b(X), D_c^b(X)) \subset D_c^b(X).$$

For the last operation $\otimes$, we need a condition of finite Tor-dimension:

Definition 5.26 ([SGA4, exp. XVII, 4.1.9]) An object $K \in D^b(X_{\text{ét}}, \mathbf{Z}/l^n)$ is *of* Tor-*dimension $\leq d$ if it satisfies the following equivalent conditions:*

(i) For all $N \in \mathbf{Z}$ and all $L \in D^b(X_{\text{ét}}, \mathbf{Z}/l^n)$ such that $H^i(L) = 0$ for $i < N$, we have $H^i(K \overset{L}{\otimes} L) = 0$ for $i < N - d$.

(ii) There exists a quasi-isomorphism $K' \to K$, with K' with flat terms which are 0 for $i < -d$.

We say that K is of finite Tor-dimension if it is of Tor-dimension $\leq d$ for a suitable d; we write

$$D_{tf}^b(X) = \{K \in D^b(X_{\text{ét}}, \mathbf{Z}/l^n) \mid K \text{ is of finite Tor-dimension}\}$$

and

$$D_{ctf}^b(X) = D_c^b(X) \cap D_{tf}^b(X).$$

We see immediately that this condition is stable under $\overset{L}{\otimes}$.

Corollary 5.27 ([SGA4$\frac{1}{2}$, Th. finitude, rem 1.7 and th. 4.3]) *The full subcategories $D_{ctf}^b(X) \subset D^b(X)$ are preserved by the six operations $f^*, Rf_*, Rf_!, Rf^!, \overset{L}{\otimes}, \text{R\underline{Hom}}$. Furthermore, the complex*

$$K_X = Rf_X^! \mathbf{Z}/l^n,$$

where $f_X : X \to S$ is the structure morphism, is dualising.

5.4.3 Passage to *l*-adic cohomology

This passage is nontrivial: a delicate point is to correctly define the categories $D^b_c(X, \mathbf{Z}_l)$ and $D^b_c(X, \mathbf{Q}_l)$, and even $D^b_c(X, \bar{\mathbf{Q}}_l)$.

We explain the problem with the first of these. We would like to define

$$D^b_c(X, \mathbf{Z}/l^n) = 2 - \varprojlim_n D^b_{ctf}(X, \mathbf{Z}/l^n) \qquad (5.4.4)$$

where, given a projective system $(\mathcal{C}_n, F_n : \mathcal{C}_n \to \mathcal{C}_{n-1})$ of categories, $2 - \varprojlim \mathcal{C}_n$ is the category whose objects are the systems

$$(X_n, u_n : F_n(X_n) \overset{\sim}{\to} X_{n-1})$$

with $X_n \in \mathcal{C}_n$, the morphisms being given "componentwise". Unfortunately, if the $\mathcal{C}_n$ are triangulated categories and the F_n are triangulated functors, the triangulated structure is in general lost after passage to $2 - \varprojlim$, due to the non-exactness of the projective limit functor (on abelian groups).

Thankfully, in the case where X is of finite type over Spec $\mathbf{Z}[1/l]$, Theorem 5.24, together with the fact that $H^i_{\text{ét}}(\operatorname{Spec}\mathbf{Z}[1/l], F)]$ is finite for all $i \in \mathbf{Z}$ and any constructible sheaf of $\mathbf{Z}/l^n$-modules F[12], implies that $\operatorname{Hom}(K, L)$ is a finite group for all $K, L \in D^b_c(X, \mathbf{Z}/l^n)$. The Mittag-Leffler condition then implies that (5.4.4) gives a well-defined triangulated category. This is the approach of Deligne in [30, 1.1.2].

The problem of finding the right definition of $D(X_{\text{ét}}, \mathbf{Z}_l)$ for arbitrary X was only resolved later by Ofer Gabber (unpublished) and independently by Torsten Ekedahl [40].

To define $D^b_c(X, \mathbf{Q}_l)$ or $D^b_c(X, \bar{\mathbf{Q}}_l)$, we start from $D^b_c(X, \mathbf{Z}_l)$ and tensor the Hom groups by $\mathbf{Q}_l$ or $\bar{\mathbf{Q}}_l$, and take the karoubian envelope if necessary. We then have:

Theorem 5.28 (cf. [30, § 1.1]) *For X varying in* $\operatorname{Sch}(\mathbf{Z}[1/l])$, *$X \mapsto D^b(X, \mathbf{Q}_l)$ admits a six operations formalism and a duality parallel to that of Corollary 5.27. (See Theorem 5.24 for the notation* Sch.)

5.4.4 Grothendieck's functional equation

We now introduce a language that will allow us to formulate the functional equation studied by Grothendieck in [CorrGS, letter of 30-9-64].

---

[12] Which essentially reduces to the finitude of the class group of $\mathbf{Z}[1/l, \mu_l]$ and to the fact that its group of units is a finite type $\mathbf{Z}$-module.

Let $X \in \mathrm{Sch}(\mathbf{F}_q)$. Suppose we are given an object $E \in D_c^b(X, \mathbf{Q}_l)$. We would like to relate the function $L(X, E)$ to the function $L(X, D_X(E))$, where D_X is the duality given above.

To explain the delicate points, we compute this last function by applying formula (5.3.9) with $Y = \mathrm{Spec}\,\mathbf{F}_q$:

$$L(X, D_X(E)) = L(\mathbf{F}_q, Rf_! D_X(E)) = L(\mathbf{F}_q, D_{\mathbf{F}_q}(Rf_* E)) = L(\mathbf{F}_q, (Rf_* E)^*)$$

where $*$ is the usual duality for graded $\mathbf{Q}_l$-vector spaces (with the Koszul rule). We thus find an expression not in terms of the cohomology with proper support on E, but in terms of ordinary cohomology.

This does not stop Grothendieck, who sets

$$R_\infty f(E) = [Rf_* E] - [Rf_! E]$$

in the Grothendieck group of $D_c^b(\mathbf{F}_q, \mathbf{Q}_l)$ (see § B.2.6). He then formulates the functional equation

$$L(X, D_X(E), t) = (-t)^{-\chi(E)} \delta(E) L(X, E, t^{-1}) A_\infty(t) \qquad (5.4.5)$$

where

$$\chi(E) = \sum_i (-1)^i \dim_{\mathbf{Q}_l} H_c^i(\bar{X}, E), \quad \delta(E) = \prod_i \det(\pi_{\mathbf{F}_q} \mid H_c^i(\bar{X}, E))^{(-1)^i}$$

and $A_\infty(t) = L(\mathbf{F}_q, R_\infty f(E), t)^{-1}$ is considered as a "corrective term at infinity". (Formula (5.4.5) is not hard to prove, by using the method of proof of (3.6.6) in Chapter 3: see the computations in (3.6.5) in the same chapter.)

Grothendieck justified this expression by giving an explicit formula for $R_\infty f(E)$ in terms of a compactification $\bar{X}$ of X:

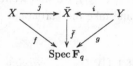

where Y is the closed complement:

$$R_\infty f(E) = [Rg_*(i^* Rj_*(E))].$$

Let us prove this formula: for all $G \in D_c^b(\bar{X}, \mathbf{Q}_l)$, we have a canonical exact triangle:

$$j_! j^* G \to G \to i_* i^* G \xrightarrow{+1} .$$

Taking $G = Rj_* E$ and using that $j^* Rj_* = Id_X$, we get

$$j_! E \to j_* E \to i_* i^* j_* E \xrightarrow{+1} .$$

Now applying $R\bar{f}_!$, we get

$$R\bar{f}_! j_! E \to R\bar{f}_! j_* E \to R\bar{f}_! i_* i^* R j_* E \xrightarrow{+1} .$$

But $\bar{f}$ is proper, so $R\bar{f}_! = R\bar{f}_*$ and the above triangle can be rewritten

$$R(\bar{f} j)_! E \to R(\bar{f} j)_* E \to R(\bar{f} i)_* i^* R j_* E \xrightarrow{+1},$$

that is,

$$R f_! E \to R f_* E \to R g_* i^* R j_* E \xrightarrow{+1} .$$

When f is proper, formula (5.4.5) becomes cleaner since $A_\infty(t) = 1$.
Suppose now that f is smooth and of pure dimension n; then

$$K_X = R f^! \mathbf{Q}_l = f^* \mathbf{Q}_l(n)[2n] = \mathbf{Q}_l(n)[2n]$$

from which we get that

$$D_X(E) = E^*(n)[2n]$$

where $E^* = \mathrm{R\underline{Hom}}(E, \mathbf{Q}_l)$ is the "naive" dual of E. In this case, (5.4.5) can
be rewritten

$$L(X, E^*, q^{-n}t) = (-t)^{-\chi(E)} \delta(E) L(X, E, t^{-1}) A_\infty(t). \qquad (5.4.6)$$

A particularly important case, which will be justified by an example below,
is where E is "weakly polarisable of weight ρ" in the sense that

$$E^* \simeq E(\rho) \qquad (5.4.7)$$

for some integer $\rho \in \mathbf{Z}$, where $E(\rho) := E \otimes \mathbf{Q}_l(\rho)$. In this case, (5.4.6) takes
the form (replacing t by $1/t$)

$$L(X, E, 1/q^{n+\rho}t) = (-t)^{\chi(E)} \delta(E) L(X, E, t) A_\infty(t^{-1}). \qquad (5.4.8)$$

We have the following generalisation of (3.6.6) "with coefficients".

Theorem 5.29 (Grothendieck, ibid.) *Suppose, in addition to the above
hypotheses, that f is proper. Thus*

- *f is smooth and proper;*
- *E is weakly polarisable of weight ρ in the sense of (5.4.7).*

Then $L(X, E, t)$ satisfies the functional equation

$$L(X, E, 1/q^{n+\rho}t) = (-t)^{\chi(E)} \delta(E) L(X, E, t)$$

with

$$\chi(E) = \sum_i (-1)^i \dim_{\mathbf{Q}_l} H^i(\bar{X}, E), \quad \delta(E) = \prod_i \det(\pi_{\mathbf{F}_q} \mid H^i(\bar{X}, E))^{(-1)^i}.$$

Furthermore, we have $\delta(E)^2 = q^{(\rho+n)\chi(E)}$, *so*

$$\delta(E) = \pm q^{\frac{(\rho+n)\chi(E)}{2}}.$$

Proof The only thing that remains to be proved is the relation between $\delta(E)$ and $\chi(E)$. But, by Poincaré duality, we may also write

$$\delta(E) = \prod_i \det(\pi_{\mathbf{F}_q} \mid H^i(\bar{X}, D_X(E)))^{(-1)^{i+1}}.$$

By replacing $D_X(E)$ with its value $E(\rho)$ and comparing to the previous expression, we easily get the desired relation.   □

5.4.5 A special case

Let $K = \mathbf{F}_q(C)$ with C smooth projective, and let $j : U \hookrightarrow C$ be a nonempty open. Suppose given a lisse $\mathbf{Q}_l$-sheaf M on U[13], and take $E = j_* M$. We then have

$$E^* = j_*(M^*),$$
$$D_C(E) = E^*(1)[2].$$

If M is weakly polarisable of weight ρ, this last formula can be rewritten

$$D_C(E) = E(\rho + 1)[2]$$

and we arrive at:

Theorem 5.30 (Grothendieck, ibid.) *With the notation and hypotheses from above, we have the same functional equation as in Theorem 5.29, with $n = 1$. Furthermore, we have a factorisation*

$$L(C, E, t) = \frac{P_1(t)}{P_0(t)P_2(t)}$$

with $P_i(t) = \det(1 - t\pi_{\mathbf{F}_q} \mid H^i(\bar{C}, j_* M))$.

---

[13] Recall that lisse = twisted-constant, cf. the note on page 111.

5.5 The theory of weights

Reference: Deligne [30].

We quote Grothendieck [CorrGS, letter of 31-10-61], in his response to Serre's letter of 26 Oct. 61 quoted in §5.3.1:

I haven't understood your question on the cohomology (of Weil, I presume) of affine varieties very well, and I also know nothing about a possible natural filtration, even for a complete variety. We can talk about it when you are at Harvard.[14]

And also (Récoltes et Semailles, vol. 5, p. 792, note 4):

I thought, wrongly, that I had introduced the filtration by weights of a motive, reflected (for all ℓ) in the corresponding filtration for the ℓ-adic realization of this motive (a filtration defined in terms of absolute values of the eigenvalues of Frobenius). In fact, Deligne reminded me that I had only worked with the notions of virtual weights (which amounted to working with virtual motives, elements of a suitable Grothendieck group . . .). It was Deligne who discovered the important fact that the virtual notion I was working with should correspond to a canonical filtration by increasing weights. This discovery (as conjectural as the conjectural theory of motives) immediately provided the key to a bona fide definition of Hodge–Deligne structures (also called mixed Hodge structures) over the complex field, as a Hodge-type transcription of structures already known on the motive and on its Hodge realization.[15]

I will now describe the theory of weights in the *l*-adic case, which is the subject of [30].

5.5.1 Weights and mixed sheaves

Definition 5.31 Let q be a prime power, and let $i \in \mathbf{Z}$. A number $\alpha \in \bar{\mathbf{Q}}$ is a *Weil q-number of weight i* if $|\rho\alpha| = q^{i/2}$ for every embedding $\rho : \bar{\mathbf{Q}} \hookrightarrow \mathbf{C}$.

---

[14] *Je n'ai pas bien compris ta question sur la cohomologie (de Weil, je présume) des variétés affines, et d'ailleurs je ne sais rien d'une éventuelle filtration naturelle, même pour une variété complète. On pourra en recauser quand tu seras à Harvard.*

[15] *Je croyais, à tort, me rappeler que j'avais introduit la filtration par les poids d'un motif, se reflétant (pour tout ℓ) en la filtration correspondante sur la réalisation ℓ-adique de ce motif (filtration définie en termes de valeurs absolues de valeurs propres de Frobenius). En fait, Deligne m'a rappelé que je n'avais travaillé qu'avec les notions de poids virtuelles (ce qui revenait à travailler avec des motifs virtuels, éléments d'un groupe de Grothendieck convenable . . .). C'est Deligne qui a découvert ce fait important que la notion virtuelle avec laquelle je travaillais devrait correspondre à une filtration canonique, par poids croissants. Cette découverte (toute aussi conjecturale que la théorie conjecturale des motifs) a fourni aussitôt la clef d'une définition en forme des structures de Hodge-Deligne (dites aussi structures de Hodge mixtes) sur le corps des complexes, comme transcription à la Hodge des structures déjà connues sur le motif et sur sa réalisation de Hodge.*

Remark 5.32 Let α be a Weil q-number of weight i. Then $\sigma\alpha$ is a Weil q-number of weight i for all $\sigma \in Gal(\bar{\mathbf{Q}}/\mathbf{Q})$. Let σ be induced by complex conjugation, via an embedding ρ. Then $\rho(\alpha \cdot \sigma\alpha)$ is a positive real number with absolute value q^i, and therefore equal to q^i. Thus q^i/α is *conjugate* to α.

Definition 5.33 Let $X \in \mathrm{Sch}(\mathbf{Z}[1/l])$ and let F be a $\mathbf{Q}_l$-adic sheaf on X.

a) F is (punctually) *pure of weight* i if, for all $x \in X_{(0)}$, the eigenvalues of $\pi_x \mid F_x$ are Weil $N(x)$-numbers of weight i (in particular, are algebraic over $\mathbf{Q}$). Here π_x denotes as usual the geometric Frobenius of x.
b) F is *mixed of weight* $\leq i$ if there exists a finite filtration of F with successive pure quotients of weight $\leq i$.

Examples 5.34 $\mathbf{Q}_l(1)$ is pure of weight -2. If $\mathcal{A}$ is an abelian scheme over X, then $V_l(\mathcal{A}) = T_l(\mathcal{A}) \otimes \mathbf{Q}_l$ is pure of weight -1 thanks to Theorem 3.36 (1). (This last result is not used in [30].)

In fact, Deligne works with a much finer notion of ι-weights and of ι-mixed sheaves, where ι is a fixed isomorphism of $\bar{\mathbf{Q}}_l$ over $\mathbf{C}$, so that mixed $\iff$ ι-mixed for all ι. Implicitly, this uses the axiom of choice! We quote him, [30, rem. 1.2.11]:

I don't claim to believe that there exist isomorphisms between $\bar{\mathbf{Q}}_l$ and $\mathbf{C}$, and these are here just for convenience of exposition. Every time we prove that a number is pure, an easy fragment of the proof would suffice to establish that it is algebraic. For the rest of the arguments, it would suffice then to consider complex embeddings of the subfield of $\bar{\mathbf{Q}}_l$ formed by the algebraic numbers, and this does not require the axiom of choice.[16]

For the correct definition of ι-weights (with values in $\mathbf{R}$) and of ι-pure and ι-mixed sheaves, see [30, 1.2.6].

We first have a striking conjecture:

Conjecture 5.35 ([30, conj. 1.2.9]) *If* $X \in \mathrm{Sch}(\mathbf{F}_p)$, *every constructible* $\mathbf{Q}_l$-*sheaf over* X *is mixed.*

Here is the main theorem.

---

[16] *Je ne prétends pas croire à l'existence d'isomorphismes entre $\bar{\mathbf{Q}}_l$ et $\mathbf{C}$, et ceux-ci ne sont qu'une commodité d'exposition. Chaque fois que nous prouverons qu'un nombre est pur, un fragment facile de la démonstration suffirait à établir qu'il est algébrique. Pour le reste des arguments, il suffirait alors de considérer les plongements complexes du sous-corps de $\bar{\mathbf{Q}}_l$ formé des nombres algébriques, et ceci ne requiert pas l'axiome du choix.*

Theorem 5.36 ([30, th. 3.3.1]) *Let* $f : X \to Y$ *be a morphism of* Sch($\mathbf{Z}[1/l]$), *and let* F *be a* $\mathbf{Q}_l$-*sheaf on* X, *mixed of weight* $\leq \rho$. *Then* $R^i f_! F$ *is mixed of weight* $\leq \rho + i$ *for all* $i \geq 0$.

Deligne then draws the following consequences for lisse sheaves.

Theorem 5.37 ([30, th. 3.4.1]) *Let* $X \in$ Sch($\mathbf{F}_q$), *and let* F *be a lisse mixed* $\mathbf{Q}_l$-*sheaf on* X. *Then*

(1) F *admits a finite increasing filtration* $W_\bullet F$, *the* weight filtration, *such that for all* $i \in \mathbf{Z}$, Gr$_i$ $F := W_i F / W_{i-1} F$ *is lisse and pure of weight* i. *This filtration is unique and strictly functorial in* F.
(2) *If* F *is pure and* X *is normal,* $F_{|\bar{X}}$ *is semisimple.*

Corollary 5.38 ([30, th. 6.1.11]) *For all* $X \in$ Sch($\mathbf{F}_q$), *write* $D_m^b(X, \mathbf{Q}_l)$ *for the thick subcategory of* $D_c^b(X, \mathbf{Q}_l)$ *formed by the objects* C *such that* $H^i(C)$ *is mixed for all* $i \in \mathbf{Z}$. *Then* D_m^b *is stable under the six operations.*

The same corollary holds for $X \in$ Sch($\mathbf{Q}$), provided we define a mixed sheaf on X as a $\mathbf{Q}_l$-sheaf arising from a mixed sheaf on $\mathcal{X}_n$ for n multiplicatively sufficiently large, where $\mathcal{X}_n$ is a model of finite type of X over $\mathbf{Z}[1/n]$ (ibid.).

The other important result is:

Theorem 5.39 ([30, th. 3.2.3]) *Let* C *be a smooth projective curve over* $\mathbf{F}_q$, $j : U \to C$ *a dense open subset, and* M *a lisse* $\mathbf{Q}_l$-*sheaf on* U, *pure of weight* ρ. *Then the eigenvalues of Frobenius acting on* $H^i(\bar{C}, j_* M)$ *are Weil* q-*numbers of weight* $\rho + i$.

Finally, we will need the following statement, which is not explicit in [30] (but see [13, 5.1.14 and 1. 5 above 5.1.9]):

Proposition 5.40 *Let* $p : \mathcal{X} \to U$ *be a smooth proper morphism, where* U *is a smooth variety of dimension* d *over* $\mathbf{F}_q$. *Let* F *be a lisse* $\mathbf{Q}_l$-*sheaf on* $\mathcal{X}$, *pure of weight* ρ. *Then* $R^i p_* F$ *is lisse and pure of weight* $\rho + i$ *for all* $i \geq 0$.

In the case $d = 0$, $F = \mathbf{Q}_l$, we recover the last of Weil's conjectures (the Riemann hypothesis) generalised from the projective smooth case to the proper smooth case [30, cor. 3.3.9].

Proof That $R^i p_* F$ is lisse follows from the fact that p is smooth and proper. Theorem 5.37 implies that it is mixed of weight $\leq \rho + i$; to conclude, it suffices to see that its dual is mixed of weight $\leq -\rho - i$. For this, we may suppose that

$\mathcal{X}$ is connected, and therefore equidimensional. We then apply the formalism (5.4.3):

$$D_U(Rp_*F) \simeq Rp_*D_{\mathcal{X}}(F)$$

which gives

$$\mathrm{R\underline{Hom}}(Rp_*F, \mathbf{Q}_l(d)[2d]) \simeq Rp_* \, \mathrm{R\underline{Hom}}(F, \mathbf{Q}_l(n+d)[2n+2d])$$

where n is the relative dimension of p, hence

$$\mathrm{R\underline{Hom}}(Rp_*F, \mathbf{Q}_l) \simeq Rp_* \, \mathrm{R\underline{Hom}}(F, \mathbf{Q}_l(n)[2n]),$$

and by taking H^{-i}:

$$\underline{\mathrm{Hom}}(R^i p_*F, \mathbf{Q}_l) \simeq R^{2n-i} p_*(\underline{\mathrm{Hom}}(F, \mathbf{Q}_l))(n).$$

(noting that $\mathrm{R\underline{Hom}}(G, \mathbf{Q}_l) = \underline{\mathrm{Hom}}(G, \mathbf{Q}_l)[0]$ for every lisse sheaf G). We conclude the proof by applying Theorem 5.37 again.   □

5.5.2 The hard Lefschetz theorem

Let X be a smooth projective variety of dimension n over a field k. Choose a smooth hyperplane section $Y \subset X$ for a suitable projective embedding i : $X \hookrightarrow \mathbf{P}^N$. By Bertini's theorem [Har2, ch. II, th. 8.18], the set of these sections defines a nonempty open subset U of an affine space: therefore Y always exists when k is infinite.

If k is finite, it may be that $U(k) = \emptyset$: in this case, we compose i with a *Veronese embedding*

$$\mathbf{P}^N = \mathbf{P}(V) \hookrightarrow \mathbf{P}(S^r(V))$$

induced by $x \mapsto x^r$. For r sufficiently large, the corresponding open subset U_r has a rational point over k (cf. [SGA7, exp. XVII]). This amounts to replacing the hyperplane sections in $\mathbf{P}^N$ by sections of hypersurfaces of degree r.

Write $L = \mathrm{cl}^1(Y) \in H^2(X)(1)$, where H is a Weil cohomology. For $i \leq n$, we may consider the homomorphism

$$H^i(X) \xrightarrow{\cdot [L]^{n-i}} H^{2n-i}(X)(n-i).$$

$$(5.5.1)$$

If $k = \mathbf{C}$ and $H = H_B$, (5.5.1) is an isomorphism; this is a consequence of Hodge theory. By using the comparison theorems from § 3.6.1, we deduce that (5.5.1) is an isomorphism if $H = H_l$ and char $k = 0$.

Over a finite field, the statement is still true, but much more difficult.

Theorem 5.41 ([30, th. 4.1.1]) *The homomorphism* (5.5.1) *is an isomorphism if* $k = \mathbf{F}_q$ *and* $H = H_l$, *with* $l \nmid q$.

On the other hand, Poincaré duality provides a perfect pairing (3.4.1). By combining these two, we get the following result.

Corollary 5.42 *For all* $i \leq n$, *the choice of L provides a perfect pairing which is Galois-equivariant and* $(-1)^i$*-symmetric:*

$$H_l^i(X) \times H_l^i(X) \to \mathbf{Q}_l(-i).$$

5.6 The completed L-function of a smooth projective variety over a global field

Reference: Serre [113].

5.6.1 The problem

Let K be a global field, and let $X \in \mathbf{V}(K)$. If $v \in \Sigma_K^f$ is a place of good reduction for X, we can define a *local factor at v* of the L-function (or zeta function) of X:

$$L_v(X, s) = \zeta(X(v), s),$$

where $X(v)$ is the special fibre at v of a smooth projective model of X over O_v, and then a partial "Hasse–Weil" L-function

$$L_0(X, s) = \prod_{\substack{v \in \Sigma_K^f \\ v \text{ good}}} L_v(X, s)$$

as we did in § 5.2. In an essentially equivalent way, the morphism $X \to \mathrm{Spec}\, K$ extends to a smooth projective morphism $\mathcal{X} \to U$, where $U = \mathrm{Spec}\, O_S$ for a suitable ring of S-integers of K if K is a number field, and U is an open subset of the smooth projective model of K if K is a function field; we then have

$$L_0(X, s) = \zeta(\mathcal{X}, s)$$

up to a finite number of Euler factors (depending on the choice of U).

For certain curves over number fields, we may prove (using Hecke characters) that $L_0(X, s)$ admits an analytic extension to $\mathbf{C}$ and a functional equation: cf. Theorem 5.2 and § 5.1.4.

In [113], Serre studied the following question: can we associate to X a "local factor" at each place $v \in \Sigma_K$, and an "exponential term" which would allow us to conjecture an elegant functional equation?

Serre responds to this question in a more precise way, by attaching local factors and an exponential term to each "cohomology group of X", which allow him to formulate a conjecture as above. Moreover, this conjecture is true in characteristic p, essentially by reduction to Theorem 5.29.

In what follows, we fix a smooth projective K-variety X of dimension n, and an integer $i \in [0, 2n]$.

5.6.2 Nonarchimedean local factors

Definition 5.43 Let $v \in \Sigma_K^f$. We set

$$L_v(K, H^i(X), s) = \det(1 - \pi_v N(v)^{-s} \mid H_l^i(X)^{I_v})^{-1}$$

where:

- l is a prime number not dividing $N(v)$.
- $H_l^i(X) = H_{\text{ét}}^i(\bar{X}, \mathbf{Q}_l)$ is the Weil cohomology associated to l (geometric l-adic cohomology).
- I_v is the inertia group at v.
- π_v is (a conjugacy class of) the geometric Frobenius element at v.

We elaborate on this definition, which of course generalises that of Artin (Definition 4.62). The absolute Galois group G_v of K_v, the completion of K at v, sits in an exact sequence

$$1 \to I_v \to G_v \to G(v) \to 1,$$

where $G(v)$ is the absolute Galois group of the residue field $\kappa(v)$. The action of G_v on $H_l^i(X)^{I_v}$ factors through $G(v)$, and the element of $G(v)$ arising in Definition 5.43 is, as usual, the inverse of the "arithmetic" Frobenius.

Lemma 5.44 *Suppose that X has good reduction at v, and let $X(v)$ be the special fibre of a smooth projective model $\mathcal{X}_v$ of X over O_v. Then*

$$L_v(K, H^i(X), s) = P_i(X(v), N(v)^{-s})^{-1}$$

where $P_i \in \mathbf{Z}[t]$ is the polynomial arising in the decomposition of the function $Z(X(v), t)$, cf. (3.6.3) in Chapter 3. In particular:

(1) $L_v(K, H^i(X), s)$ *does not depend on the choice of l;*
(2) $\prod_{i=0}^{2n} L_v(K, H^i(X), s)^{(-1)^i} = \zeta(X(v), s)$.

Furthermore, the infinite product $\prod_v L_v(K, H^i(X), s)$, extended to the places of good reduction of X, converges absolutely for $\Re(s) > 1 + i/2$.

Proof The first statement follows from the smooth and proper base change theorem, cf. § 3.6.1: this includes the fact that (under the hypothesis of good reduction) I_v acts trivially on $H_l^i(X)$. The integrality of the P_i, their independence of l and convergence of the resulting product all follow from the Riemann hypothesis [26], cf. Lemma 3.68. The identity in (2) is clear. □

5.6.3 The case of bad reduction

What happens when X does not have good reduction at v? The independence of l is then a conjecture.

Conjecture 5.45 (Serre [113, conj. C_5 and C_6]) *The polynomial*

$$\det(1 - \pi_v t \mid H_l^i(X)^{I_v})$$

is independent of the choice of l and has integer coefficients; its inverse roots are Weil numbers of weights between 0 and i.

Theorem 5.46 *Conjecture 5.45 holds in the following cases:*

(i) *If $i \in \{0, 1, 2n - 1, 2n\}$.*
(ii) *If char $K > 0$.*

Proof (i) is trivial for $i = 0$; it is a theorem of Grothendieck for $i = 1$: [SGA7, exp. IX, th. 4.3 and cor. 4.4]. The other cases then follow from Poincaré duality.

(To apply the results of Grothendieck, we observe that there is an isomorphism $H_l^1(X)(1) \xrightarrow{\sim} T_l(\text{Pic}^0(X))$, where $\text{Pic}^0(X)$ is the Picard variety of X: this follows from Theorem 3.12 and from the Kummer exact sequence.)

The case (ii) is due to Tomohide Terasoma [132]. □

5.6.4 The weight-monodromy conjecture

Conjecture 5.45 is linked to the weight-monodromy conjecture, which can be summarised by the slogan:

On $H_l^i(X)$, the weight filtration coincides with the monodromy filtration.

More precisely, one defines a certain $\text{Gal}(\bar{K}/K)$-invariant filtration $M_* H_l^i(X)$ of $H_l^i(X)$: the *monodromy filtration* [30, 1.7]. It is subject to *Grothendieck's l-adic monodromy theorem* (see Theorem 5.48 below).

Definition 5.47 Let Γ be a profinite group, l be a prime number, and $\rho : \Gamma \to GL_n(\mathbf{Q}_l)$ be a continuous representation. We say that ρ is *unipotent* if the endomorphism $\rho(g) - 1$ is nilpotent for all $g \in \Gamma$. We say that ρ is *quasi-unipotent* if there exists an open subgroup $U \subset \Gamma$ such that the restriction of ρ to U is unipotent.

Theorem 5.48 ([SGA7, exp. I]) *Let K be a field complete for a discrete valuation, with finite residue field k of characteristic p, let $\rho : G \to GL_n(\mathbf{Q}_l)$ be a continuous representation, where $G = \mathrm{Gal}(K_s/K)$ and let $l \neq p$. Then the restriction of ρ to the inertia group I is quasi-unipotent.*

See Exercise 5.51 for a proof, and [SGA7, loc. cit.] for others (with weaker hypotheses on k).

This theorem implies that the action of I_v on $\mathrm{Gr}_j^M H_l^i(X)$ factors through a finite quotient; if F is a lift of the geometric Frobenius in $\mathrm{Gal}(\bar{K}_v/K_v)$, its action on $\mathrm{Gr}_j^M H_l^i(X)$ is then well-defined up to a root of unity. This allows us to formulate:

Conjecture 5.49 *For all $j \in \mathbf{Z}$, the eigenvalues of Frobenius acting on $\mathrm{Gr}_j^M H_l^i(X)$ are Weil numbers of weight $i + j$.*

For $K = \mathbf{F}_q(C)$ for a curve C, the conjecture is proved by Deligne in [30, 1.8.4]; Terasoma's proof of Theorem 5.46 (ii) relies on this result.

Following Takeshi Saito [100, cor. 0.6], Conjecture 5.45 follows from Conjecture 5.49 and the algebraicity of the ith Künneth projector of X (see § 6.9 for this notion). This allows him to prove Conjecture 5.45 for *surfaces over number fields*: indeed, it is known in this case that all Künneth projectors are algebraic (Theorem 6.31(2)), and the proof of the weight-monodromy conjecture is due to Rapoport–Zink [93, Satz 2.13].

Moreover, Morihiko Saito showed that (over a number field) Conjecture 5.49 follows from the standard conjectures of Grothendieck [99]. Finally, Peter Scholze proved it in several new cases [105].

Exercise 5.50 Let l be a prime number; let K be a finite extension of $\mathbf{Q}_l$ and let O_K be its ring of integers.

(a) Let $x \in 1 + l^n O_K$. Show that $x^l \in 1 + l^{n+1} O_K$.
(b) Let $x \in 1 + l O_K$ and $a \in \mathbf{Z}_l$. Let (a_r) be a sequence of natural numbers converging l-adically towards a. Show that the sequence x^{a_n} converges in $1 + l O_K$ towards a limit x^a which depends only on a.

(c) Prove the identities

$$(xy)^a = x^a y^a, \quad (x^a)^b = x^{ab}.$$

(d) Let m be an integer prime to l. Let k be the residue field of K. Show that the reduction homomorphism $\mu_m(K) \to \mu_m(k)$ is bijective. (Use Hensel's lemma.)

(e) Let $\zeta \neq 1$ be a root of unity of K. Show that we have $\zeta \notin 1 + lO_K$ if $l > 2$, and that $\zeta \notin 1 + 4O_K$ if $l = 2$. (First treat the case where ζ has order prime to l, by using (d). Otherwise, reduce to the case where ζ has order l if $l > 2$ and order 4 if $l = 2$; treat the case $l = 2$, $\zeta = -1$ separately.)

(f) Let $n \geq 1$. Show that $U = 1 + l M_n(\mathbf{Z}_l)$ is a subgroup of $GL_n(\mathbf{Z}_l)$, and is a pro-l-group. (Let $U^{(r)} = 1 + l^r M_n(\mathbf{Z}_l)$, $r \geq 1$. Filter U by the $U^{(r)}$, and construct an isomorphism of groups

$$U^{(r)}/U^{(r+1)} \xrightarrow{\sim} M_n(\mathbf{F}_l)$$

for all r.)

Exercise 5.51 (The l-adic monodromy theorem) Some knowledge of the ramification theory of local fields will be necessary to do this exercise; see [Ser2, ch. IV, §§ 1 and 2]. The exercise follows the proof of [114, Prop. in the appendix]. One may use Exercises 5.10 and 5.50.

(a) Let ρ be as in Theorem 5.48, and let V be its representation space. Show that G leaves a lattice of V invariant. (Let L_0 be an arbitrary lattice of V: show that $\rho(G) \cdot L_0$ is compact, and therefore a lattice.)

(b) Deduce that we may suppose that ρ takes values in $GL_n(\mathbf{Z}_l)$.

(c) Show that it suffices to prove the theorem with G replaced by an open subgroup, that is, K by a finite extension.

(d) Deduce that for all $\sigma \in G$ we may suppose that we have $\rho(\sigma) \equiv 1_n$ (mod $l^2 M_n(\mathbf{Z}_l)$) where 1_n is the identity matrix (consider the composition $G \xrightarrow{\rho} GL_n(\mathbf{Z}_l) \to GL_n(\mathbf{Z}/l^2)$). We will then show that $\rho_{|I}$ is unipotent.

(e) Show that $\rho(P) = 1$, where P is the wild inertia group of K. (Use Exercise 5.50 (f).)

(f) Show that ρ even factors through the quotient $I_m^l = \mathbf{Z}_l(1)$ of I_m.

(g) Let $s \in I_m^l$, and let $\lambda \in L$ be an eigenvalue of $\rho(s)$ in a finite extension L of $\mathbf{Q}_l$. Show that $\lambda \in 1 + l^2 O_L$.

(h) Let q be the cardinality of the residue field of K. Show that λ^q is also an eigenvalue of $\rho(s)$. (Use Exercise 5.10 (a) and the fact that the homomorphism of [Ser2, ch. IV, § 2, cor. 1] is Galois-equivariant.)

(i) Deduce that λ^{q^i} is an eigenvalue of $\rho(s)$ for all $i \geq 0$, and then that λ is a root of unity.

(j) Deduce that $\lambda = 1$. (Use (g) and Exercise 5.50 (d).)

(k) Conclude that $\rho(s)$ is indeed unipotent.

5.6.5 Local archimedean factors

Now let $v \in \Sigma_K^\infty$. To define a local factor $\Gamma_v(K, H^i(X), v)$, we use the *Hodge decomposition*

$$H^i(X_v(\mathbf{C}), \mathbf{C}) = \bigoplus_{p+q=i} H^{p,q}$$

where $X_v = X \otimes_K K_v$.

We start with the complex case:

Definition 5.52 Suppose $K_v = \mathbf{C}$, and set

$$\Gamma_v(K, H^i(X), s) = \prod_{p,q} \Gamma_{\mathbf{C}}(s - \inf(p,q))^{h(p,q)}, \quad h(p,q) = \dim H^{p,q}.$$

The real case is more interesting: complex conjugation

$$\pi_v \in \mathrm{Gal}(\bar{K}_v/K_v) \simeq \mathbf{Z}/2$$

acts on $X_v(\mathbf{C})$ (we identify $\bar{K}_v$ with $\mathbf{C}$), and therefore on $H^i(X_v(\mathbf{C}), \mathbf{C})$ by sending $H^{p,q}$ to $H^{q,p}$.

Definition 5.53 Suppose $K_v = \mathbf{R}$.

a) If i is odd, we set

$$\Gamma_v(K, H^i(X), s) = \prod_{p<q} \Gamma_{\mathbf{C}}(s - p)^{h(p,q)}.$$

b) If $i = 2j$ is even, we set

$$\Gamma_v(K, H^i(X), s) = \prod_{p<q} \Gamma_{\mathbf{C}}(s-p)^{h(p,q)} \Gamma_{\mathbf{R}}(s-j)^{h(j,+)} \Gamma_{\mathbf{R}}(s-j+1)^{h(j,-)},$$

where $h(j, \varepsilon)$ is the dimension of the eigenspace of the eigenvalue $(-1)^j \varepsilon$ for the action of π_v on $H^{j,j}$.[17]

---

[17] Note that the function $\Gamma_{\mathbf{C}}(s)$ used here is twice the one used by Serre in [113], hence $\Gamma_v(K, H^i(X), s)$ differs from Serre's Gamma factor by a power of 2. Of course this does not change Conjecture 5.57 below.

5.6.6 The exponential factor

The definition of the exponential factor involves the *conductor of $H^i(X)$*: this is an effective divisor

$$\mathfrak{f} = \sum_{v \in \Sigma_K^f} f(v)v$$

where $f(v) = \varepsilon_v + \delta_v$; these integers are defined as follows. Firstly,

$$\varepsilon_v = \dim H_l^i(X) - \dim H_l^i(X)^{I_v}.$$

To define δ_v, we replace the representation $H_l^i(X)$ of G_v by its *semi-simplification* $H_l^i(X)^{ss}$: the l-adic monodromy theorem implies that the action of I_v on $H_l^i(X)^{ss}$ factors through a finite quotient. We can then define its *Swan conductor* (a variant of the Artin conductor, [94, p. 130] or [113, § 2.1]): δ_v is its exponent.

We also define the discriminant of K by

$$D = \begin{cases} |d_{K/\mathbf{Q}}| & \text{if char } K = 0, \\ q^{2g-2} & \text{if } K = \mathbf{F}_q(C), \end{cases}$$

where, in the second case, C is the smooth projective model of K, g is its genus, and $\mathbf{F}_q$ is its field of constants.

Definition 5.54 $A = N(\mathfrak{f})D^{B_i}$ where $B_i = \dim_{\mathbf{Q}_l} H_l^i(X)$.

5.6.7 The completed L-function

Definition 5.55 (Serre [113], up to notation) We set

$$L(K, H^i(X), s) = \prod_{v \in \Sigma_K^f} L_v(K, H^i(X), s),$$

$$\Lambda_\infty(K, H^i(X), s) = A^{s/2} \prod_{v \in \Sigma_K^\infty} \Gamma_v(K, H^i(X), s),$$

$$\Lambda(K, H^i(X), s) = \Lambda_\infty(K, H^i(X), s)L(K, H^i(X), s).$$

Remark 5.56 The exponential term $A^{s/2}$ decomposes into a product of local factors: T. Saito suggested to me that it could be interesting to incorporate them into the corresponding local factors $L_v(K, H^i(X), s)$.

5.6.8 The functional equation

Conjecture 5.57 (Serre, ibid.) *The function* $\Lambda(K, H^i(X), s)$ *admits a mero-morphic extension to the whole complex plane, and a functional equation*

$$\Lambda(K, H^i(X), s) = w\Lambda(K, H^i(X), i + 1 - s), \quad w = \pm 1.$$

For X an elliptic curve, this conjecture (in a non-cohomological form) goes back to Weil [146].

Theorem 5.58 (Grothendieck, Deligne) *Conjecture 5.57 holds if* char $K > 0$. *Moreover, if* $K = \mathbf{F}_q(C)$ *as above, we have*

$$\Lambda(K, H^i(X), s) = A^{s/2} \frac{P_1(q^{-s})}{P_0(q^{-s}) P_2(q^{-s})}$$

with $P_0, P_1, P_2 \in \mathbf{Z}[t]$, *where the inverse roots of* P_j *are Weil q-numbers of weight* $i + j$.

Proof The smooth projective morphism $X \to \operatorname{Spec} K$ extends to a smooth projective morphism $p : \mathcal{X} \to U$, where U is a suitable open subset of C; if $j_U : U \hookrightarrow C$ is the corresponding immersion, we have

$$j_* H_l^i(X) = (j_U)_* R^i p_* \mathbf{Q}_l$$

where $j : \operatorname{Spec} K \to C$ is the inclusion of the generic point.

Thus, $j_* H_l^i(X)$ is a mixed $\mathbf{Q}_l$-sheaf on C. I claim that up to an exponential term, we have:

$$\Lambda(K, H^i(X), s) = L(C, j_* H_l^i(X), q^{-s}).$$

Indeed, consider the local factor of the right-hand term at $x \in C_{(0)}$:

$$L(\kappa(x), i_x^* j_* H_l^i(X), q^{-s}) = L(\kappa(x), i_x^* (j_x)_* H_l^i(X), q^{-s})$$

where j_x is the local inclusion $\operatorname{Spec} K \to \operatorname{Spec} \mathcal{O}_{C,x}$. It follows from the Galois description of the direct image under such an open immersion (for example, [Mil, ch. II, ex. 3.15]) that

$$(j_x)_* H_l^i(X) = H_l^i(X)^{I_x}$$

as a module over the absolute Galois group of K. Then

$$L(\kappa(x), i_x^* (j_x)_* H_l^i(X), q^{-s}) = L_x(K, H^i(X), s)$$

in the sense of Definition 5.43.

By Proposition 5.40, $R^i p_* \mathbf{Q}_l$ is lisse and pure of weight i. Corollary 5.42 implies that the sheaf $H^i_l(X)$ is "weakly polarisable of weight i"; up to restricting U[18], this statement extends to the sheaf $R^i p_* \mathbf{Q}_l$. Applying Theorem 5.30, we then find a functional equation of the form

$$L(C, j_* H^i_l(X), q^{i+1-s}) = \pm(-q^{-s})^\chi q^{\frac{(i+1)\chi}{2}} L(C, j_* H^i_l(X), q^s) \quad (5.6.1)$$

where $\chi = \chi(C, j_* H^i_l(X))$, and the announced factorisation *(a priori*, with coefficients in $\mathbf{Q}_l$).

Moreover, Theorem 5.39 implies that the inverse roots of P_j are Weil numbers of weight $i + j$. Next, Theorem 5.46 implies that $L(C, j_* H^i_l(X), t) \in \mathbf{Z}[[t]]$: this is then a rational function with rational coefficients (cf. the proof of Theorem 3.65). The same reasoning as in the proof of Lemma 3.68 then shows that the P_j are pairwise coprime with integer coefficients and independent of the choice of l. It remains to deduce from (5.6.1) the functional equation given in the statement of Conjecture 5.57. This follows from the *Grothendieck–Ogg–Šafarevič formula* [94, th. 1]:

$$\chi(C, j_* H^i_l(X)) = (2 - 2g)B_i - \deg \mathfrak{f}$$

(cf. § 5.6.6 for $\mathfrak{f}$). An ancestor of this formula is due (as always...) to Weil: [Ser2, ch. VI, § 4].

(This argument is not quite complete: the Grothendieck–Ogg–Šafarevič formula is valid for constructible $\mathbf{Z}/l^n$-sheaves; we must then apply it to $j_* H^i_{\text{ét}}(\bar{X}, \mathbf{Z}/l^n)$ for all $n \geq 1$ and then pass to the limit. The situation is comparable to the one described in § 5.3.8.) □

For the constant of the functional equation, I refer to Deligne's exposé [24] which relies essentially on Theorem 4.74 and the weight-monodromy conjecture of § 5.6.4.

---

[18] To provide a good polarisation with respect to p.

6

Motives

References: Kleiman [72], Scholl [104], André [And].
  We fix a base field k.

6.1 The issue

We quote Serre [115]:

The situation described above is not quite satisfactory. One has too many cohomology groups that are not sufficiently interrelated – in spite of the compatibility isomorphisms. For example, if X and Y are two (smooth, projective) varieties, and $f : H^i_{\text{ét}}(X, \mathbf{Q}_\ell) \to H^i_{\text{ét}}(Y, \mathbf{Q}_\ell)$ is a $\mathbf{Q}_\ell$-linear map, where ℓ is a fixed prime, it is not possible in general to infer from f an analogous map for ℓ'-adic cohomology, where ℓ' is another prime number. However, one feels that this is possible for certain f, those that are "motivated" (for example those which arise from a morphism from Y to X, or more generally from an algebraic correspondence between X and Y). Still, one needs to know what it means for f to be "motivated"![1]

The compatibility isomorphisms Serre talks about are those from § 3.6.1 as well as an analogous isomorphism between Betti cohomology and de Rham cohomology: these only exist in characteristic zero. In characteristic p, the situation is more mysterious: there are no comparison isomorphisms linking the

---

[1] *La situation décrite ci-dessus n'est pas tout à fait satisfaisante. On dispose de trop de groupes de cohomologie qui ne sont pas suffisamment liés entre eux – malgré les isomorphismes de compatibilité. Par exemple, si X et Y sont deux variétés (projectives, lisses), et*
$f : H^i_{\text{ét}}(X, \mathbf{Q}_\ell) \to H^i_{\text{ét}}(Y, \mathbf{Q}_\ell)$ *une application $\mathbf{Q}_\ell$-linéaire, où ℓ est un nombre premier fixé, il n'est pas possible en général de déduire de f une application analogue pour la cohomologie ℓ'-adique, où ℓ' est un autre nombre premier. Pourtant, on a le sentiment que c'est possible pour certains f, ceux qui sont «motivés» (par exemple ceux qui proviennent d'un morphisme de Y dans X, ou plus généralement d'une correspondance algébrique entre X et Y). Encore faut-il savoir ce que «motivé» veut dire!*

different *l*-adic cohomologies. However, the characteristic polynomial of the action of Frobenius on *l*-adic cohomology has integer coefficients independent of *l*...

One might imagine that there exists an initial Weil cohomology, with coefficients in $\mathbf{Q}$, with maps to all the *l*-adic cohomologies: this is impossible, by Serre's observation, described in § 3.5.9. What then?

Grothendieck's idea was to find an initial Weil cohomology, not with values in some category of vector spaces over a field, since this is impossible, but in some suitable abelian category. This category was to be the *category of motives*, and the universal Weil cohomology would be *motivic cohomology*.[2]

Grothendieck's idea for giving a sense to "motivated" is, as suggested by Serre, to start from algebraic correspondences, the common point between the various Weil cohomologies via cycle class maps. This already defines an additive category whose objects are the smooth projective varieties; by refining this definition by purely categorical procedures, we arrive at a *rigid* symmetric monoidal category (Theorem 6.19): the category of pure motives $\mathcal{M}_\sim(k, F)$.

We have implicitly made a choice of a coefficient field F and an adequate equivalence relation $\sim$ (§ 6.2). If we choose $\sim$ to be numerical equivalence, then the category $\mathcal{M}_\sim(k, F)$ is *semisimple abelian* (Theorem 6.20). If k is finite, it follows from the last of Weil's conjectures (Deligne) that $\mathcal{M}_\sim(k, F)$ admits a *gradation by weights* (Theorem 6.33 and Proposition 6.41) and part of Grothendieck's program is achieved. In particular, every simple motive has a well-determined weight i, its zeta function is a polynomial or the inverse of a polynomial, depending on the parity of i, and its inverse roots are Weil numbers of weight i (Theorem 6.50). As a particular case, we will recover Weil's theorem on Artin's conjecture in characteristic p in § 6.15.

Unfortunately, some open problems limit the applicability of Grothendieck's definition: the *standard conjectures*, already mentioned in § 3.6.1. For example, we do not know if the projectors that define a grading of a Weil cohomology (the "Künneth components of the diagonal") are always represented by algebraic cycles. To approach L-functions over number fields, Deligne then introduced a variant of Grothendieck's category of motives: the Homs of this category consist not of algebraic cycles, but of "absolute Hodge cycles" [29, DMOS], and the standard conjectures hold thanks to the hard Lefschetz theorem and the polarisability of the Hodge structures carried by the Betti cohomology of smooth projective complex varieties. This was greatly

---

[2] This terminology was used by Lichtenbaum, and then Friedlander, Suslin and Voevodsky, in a very different sense (Hom groups in the category of triangulated motives): one should take care not to confuse the two notions.

refined by Yves André [2], who replaced absolute Hodge cycles by "motivated cycles". I will not address these theories, nor the most exciting aspects of the theory of motives, which extend far beyond the problem of the Weil conjectures: Tannaka duality and motivic Galois groups [Saa, DMOS, 2].

6.2 Adequate equivalence relations

Definition 6.1 Let $V(k)$ be the category of smooth projective k-varieties, and let F be a commutative ring. An equivalence relation $\sim$ on algebraic cycles with coefficients in F is *adequate* if it is F-linear and if it satisfies:

(i) **Moving lemma** Let $\alpha, \beta \in Z^*(X, F) = Z^*(X) \otimes_{\mathbf{Z}} F$ for $X \in V(k)$. Then there exist $\beta' \sim \beta$ such that α and β' intersect properly (that is to say, with the correct codimensions).

(ii) Let $X, Y \in V(k)$, and let $\alpha \in Z^*(X)$, $\beta \in Z^*(X \times Y)$ be such that $\alpha \times Y$ and β intersect properly. Then $a \sim 0$ implies that $(p_Y)_*(\beta \cdot (\alpha \times Y)) \sim 0$.

We say that $(F, \sim)$ is an *adequate pair*.

Given an adequate pair $(F, \sim)$ and an $X \in V(k)$, we write

$$A_{\sim}^*(X, F) = Z^*(X, F)/\sim \quad \text{and} \quad A_*^{\sim}(X, F) = Z_*(X, F)/\sim .$$

We can do intersection theory on $A_{\sim}^*(X, F)$.

Examples 6.2 Here are the main adequate equivalence relations:

$$\sim_{\text{rat}} \geq \sim_{\text{alg}} \geq \sim_{\text{tnil}} \geq \sim_H \geq \sim_{\text{num}}$$

respectively, rational equivalence, algebraic equivalence, Voevodsky's smash-nilpotence equivalence, homological equivalence relative to a Weil cohomology H (when $F = \mathbf{Q}$), and numerical equivalence. The symbol $\geq$ denotes implication between these equivalences. Rational equivalence is the finest adequate equivalence, and numerical equivalence is the coarsest when F is a field. For more details, one may consult Fulton [Ful, ch. 19] or André [And, ch. 3]. We extract the following result from the second reference for later use.

Lemma 6.3 $\sim_{\text{rat}}$ *is the finest adequate equivalence relation.*

Proof (following [And, 3.2.2.1]) Let $\sim$ be an adequate equivalence relation for algebraic cycles with coefficients in F. Condition (ii) reduces us to proving that $[0] \sim [\infty]$ on $\mathbf{P}^1$. By condition (i), there exists a cycle $\sum n_i[x_i] \sim [1]$, $n_i \in F$, such that $\sum n_i[x_i] \cdot [1]$ is well-defined, i.e. $x_i \neq 1$. Apply (ii) by taking the graph of the rational function $1 - \prod \left(\frac{x - x_i}{1 - x_i} \right)^{m_i}$ ($m_i > 0$) for β and setting $\alpha = \sum n_i [x_i] - [1]$: we get that $mn[1] \sim m[0]$, where $m = \sum m_i, n = \sum n_i$; since the m_i are arbitrary, we conclude that $n[1] \sim [0]$. By applying the automorphism $x \mapsto 1/x$ (and condition (ii) once again), we get $n[1] \sim [\infty]$, and hence $[0] \sim [\infty]$. $\quad\square$

So as not to make our notation cumbersome, we write A^*_{num} in place of $A^*_{\sim_{\text{num}}}$, etc. We then have

Theorem 6.4

(1) *The groups of cycles (with integer coefficients) modulo numerical equivalence are free $\mathbf{Z}$-modules of finite type.*
(2) *For any commutative ring F without $\mathbf{Z}$-torsion and for all $X \in \mathbf{V}(k)$, the obvious surjection*

$$A^*_{\text{num}}(X, \mathbf{Z}) \otimes_{\mathbf{Z}} F \to A^*_{\text{num}}(X, F)$$

is bijective.

Proof (following [And, 3.4.6]) (1) Let H be a Weil cohomology with coefficient field K. For X smooth projective of dimension d, and for $r \geq 0$, let $A^{d-r}_H(X)$ be the quotient of $Z^{d-r}(X, \mathbf{Z})$ by the homological equivalence associated to H: it is isomorphic to the image of the cycle class map cl in the finite-dimensional K-vector space $H^{2(d-r)}(X)(d-r)$. Let $(\beta_1, \ldots, \beta_n)$ be a maximal system of elements of $A^{d-r}_H(X)$ such that $\text{cl}(\beta_1), \ldots, \text{cl}(\beta_n)$ are K-linearly independent. Then the homomorphism from $A^r_{\text{num}}(X)$ to $\mathbf{Z}^n$ induced by

$$\alpha \mapsto (\deg(\alpha \beta_1), \ldots, \deg(\alpha \beta_n))$$

is *injective*: indeed, if α is in its kernel and if $\beta \in A^{d-r}_H(X)$, then $\text{cl}(\beta)$ is a K-linear combination of the $\text{cl}(\beta_i)$, so that $\deg(\alpha\beta) = 0$ by compatibility of the intersection product and the Poincaré pairing (Definition 3.41 (viii)). This concludes the proof, since $\mathbf{Z}$ is Noetherian.

(2) We may suppose that X is irreducible. Write $A = A^r_{\text{num}}(X, \mathbf{Z})$ and $B = A^{d-r}_{\text{num}}(X, \mathbf{Z})$, where $d = \dim X$. Then we have to show that the pairing

$$A \otimes_{\mathbf{Z}} F \times B \otimes_{\mathbf{Z}} F \to F$$

induced by the intersection product is nondegenerate. Since F is flat over $\mathbf{Z}$, the injection $A \hookrightarrow \text{Hom}_{\mathbf{Z}}(B, \mathbf{Z})$ induces an injection

$$A \otimes_{\mathbf{Z}} F \hookrightarrow \text{Hom}_{\mathbf{Z}}(B, \mathbf{Z}) \otimes_{\mathbf{Z}} F.$$

But the second term gets sent bijectively to $\text{Hom}_F(B \otimes_{\mathbf{Z}} F, F)$ since B is $\mathbf{Z}$-free and of finite type by (1).   □

Remark 6.5 There is an analogy between the proof of (1) and the proof of the finitude of the integral closure in a separable extension, cf. [Ser2, ch. I, prop. 8].

6.3 The category of correspondences

Definition 6.6 Let $\sim$ be an adequate equivalence relation. The category $\text{Corr}_\sim(k, F)$ is given by:

Objects: $[X]$, $X \in \mathbf{V}(k)$.
Morphisms: $\text{Corr}_\sim(k, F)([X], [Y]) = \text{Corr}_\sim^0(X, Y)$, where $\text{Corr}_\sim^0$ is the group of correspondences modulo $\sim$, cf. Definition 3.52 in the case of $\sim_{\text{rat}}$.

Composition of correspondences is defined as in this definition.

Definition 6.7 We have a contravariant functor from $\mathbf{V}(k)$ to $\text{Corr}_\sim(k, F)$:

$$\mathbf{V}(k)^{\text{op}} \to \text{Corr}_\sim(k, F),$$
$$X \mapsto [X],$$
$$f \mapsto {}^t[\Gamma_f].$$

The category $\mathbf{V}(k)$ is unital symmetric monoidal (Definition A.38) for the product of varieties (with unit: $\text{Spec}\, k$), and we have:

Lemma 6.8 *The functor* [ ] *commutes with finite coproducts; the category* $\text{Corr}_\sim(k, F)$ *is F-linear. Moreover, there exists a canonical unital symmetric monoidal structure which is F-linear on* $\text{Corr}_\sim(k, F)$ *such that the above functor is symmetric monoidal.*

Proof By construction, $\text{Corr}_\sim(k, F)$ is an F-category (its Homs are F-modules and composition is F-bilinear); given $X, Y \in \mathbf{V}(k)$, we check immediately that $[X] \oplus [Y]$ is represented by $[X \coprod Y]$ and that $[\emptyset]$ represents the zero object. For the second statement, we define $[X] \otimes [Y] = [X \times Y]$, and the tensor product of correspondences is given by product of cycles:

$$\text{Corr}_\sim(X, Y) \otimes \text{Corr}_\sim(X', Y') \to \text{Corr}_\sim(X \times X', Y \times Y').$$

The verification of the axioms follows easily.   □

6.4 Pure effective motives

Definition 6.9 We retain the above notation. The category $\mathcal{M}^{\mathrm{eff}}_\sim(k, F)$ of *effective motives* (with coefficients in F, relative to $\sim$) is the Karoubi envelope of $\mathrm{Corr}_\sim(k, F)$ (Definition A.5). We abbreviate this notation as $\mathcal{M}^{\mathrm{eff}}_\sim(k)$ or $\mathcal{M}^{\mathrm{eff}}_\sim$ when there is no risk of ambiguity.

From Lemma A.9 and Proposition A.33, the symmetric monoidal F-linear structure of $\mathrm{Corr}_\sim(k, F)$ extends canonically to $\mathcal{M}^{\mathrm{eff}}_\sim(k, F)$. We write

$$[X] \mapsto h(X)$$

for the canonical functor $\mathrm{Corr}_\sim(k, F) \to \mathcal{M}^{\mathrm{eff}}_\sim(k, F)$. We write $h_\sim$ instead of h when precision requires it.

Following the proof of Lemma 6.3, $[0] = [\infty]$ in $A^\sim_0(\mathbf{P}^1, F)$; by applying a suitable homography $z \mapsto \frac{az+b}{cz+d}$, we deduce that $[x] = [y]$ in $A^\sim_0(\mathbf{P}^1, F)$ for all $x, y \in \mathbf{P}^1(k)$. Thus, the class of the projector

$$p : \mathbf{P}^1 \to \mathrm{Spec}\, k \xrightarrow{i_x} \mathbf{P}^1$$

in $\mathrm{Corr}^0_\sim(\mathbf{P}^1, F)$ does not depend on the choice of $x \in \mathbf{P}^1(k)$.

Definition 6.10 The *Lefschetz motive* $\mathbb{L}$ is the image of the projector $1 - p$. We have

$$h(\mathbf{P}^1) = \mathbf{1} \oplus \mathbb{L}.$$

We use the abbreviation

$$M(-n) = M \otimes \mathbb{L}^{\otimes n}$$

for $M \in \mathcal{M}^{\mathrm{eff}}_\sim(k, F)$ an effective motive and $n \geq 0$ an integer.

Proposition 6.11 $\mathbb{L}$ *is quasi-invertible (Definition A.28).*

Proof For every $X \in \mathbf{V}(k)$, the projectors p and $1 - p$ induce the *projective line formula*

$$A^n_\sim(X \times \mathbf{P}^1, F) \simeq A^n_\sim(X, F) \oplus A^{n-1}_\sim(X, F) \tag{6.4.1}$$

(reduce to the case of rational equivalence, cf. Lemma 6.3, then apply [Ful, th. 3.3]). By iterating once and following the direction of the arrows, one sees easily that this provides an isomorphism

$$\mathcal{M}^{\mathrm{eff}}_\sim(h(X), h(Y)) \xrightarrow{\sim} \mathcal{M}^{\mathrm{eff}}_\sim(h(X)(-1), h(Y)(-1))$$

induced by $- \otimes \mathbb{L}$. (See [81, § 8, Lemma] for details.) $\qquad\square$

We also have:

Proposition 6.12 *For all $X \in \mathbf{V}(k)$ and all $n \geq 0$, one has natural isomorphisms*

$$A^n_\sim(X, F) \simeq \mathcal{M}^{\mathrm{eff}}_\sim(\mathbb{L}^n, h(X)), \quad A^\sim_n(X, F) \simeq \mathcal{M}^{\mathrm{eff}}_\sim(h(X), \mathbb{L}^n).$$

Furthermore, the composition

$$\mathcal{M}^{\mathrm{eff}}_\sim(\mathbb{L}^n, h(X)) \times \mathcal{M}^{\mathrm{eff}}_\sim(h(X), \mathbb{L}^n) \to \mathcal{M}^{\mathrm{eff}}_\sim(\mathbb{L}^n, \mathbb{L}^n) \xleftarrow{\sim} F$$

can be identified with the intersection product

$$A^n_\sim(X, F) \times A^\sim_n(X, F) \to F.$$

Proof This follows once again from the formula (6.4.1), by iteration. □

6.5 Pure motives

Definition 6.13 We write

$$\mathcal{M}_\sim(k, F) = \mathcal{M}^{\mathrm{eff}}_\sim(k, F)[\mathbb{L}^{-1}]$$

cf. Definition A.30.

Taking Proposition 6.11 into account, Theorem A.32 gives:

Theorem 6.14 *The canonical functor $\mathcal{M}^{\mathrm{eff}}_\sim(k, F) \to \mathcal{M}_\sim(k, F)$ is fully faithful; the symmetric monoidal structure of $\mathcal{M}^{\mathrm{eff}}_\sim(k, F)$ extends to $\mathcal{M}_\sim(k, F)$.* □

Definition 6.15 We let $\mathbb{T} = \mathbb{L}^{-1}$ be a quasi-inverse of $\mathbb{L}$: this is the *Tate motive*. For $M \in \mathcal{M}_\sim(k, F)$ and $n \in \mathbf{Z}$, we write

$$M(n) = \begin{cases} M \otimes \mathbb{L}^{\otimes -n} & \text{if } n \leq 0, \\ M \otimes \mathbb{T}^{\otimes n} & \text{if } n \geq 0. \end{cases}$$

To summarise, we have constructed a chain of categories and symmetric monoidal functors

$$\mathbf{V}(k)^{\mathrm{op}} \longrightarrow \mathrm{Corr}_\sim(k, F) \xrightarrow{\ \natural\ } \mathcal{M}^{\mathrm{eff}}_\sim(k, F) \xrightarrow{\ \mathbb{L}^{-1}\ } \mathcal{M}_\sim(k, F)$$

$$X \quad \mapsto \quad [X] \quad \mapsto \quad h(X) \quad \mapsto \quad h(X)$$

$$f \quad \mapsto \quad [{}^t\Gamma_f]$$

and these last two functors are fully faithful.

Remark 6.16 So far we have introduced the classical construction of pure motives, as presented by Grothendieck. In [57], Uwe Jannsen introduced a new, direct construction $\mathcal{M}_\sim(k, F)$ as the category of triples (X, p, n) where $X \in \mathbf{V}(k)$, p is a projector of X and $n \in \mathbf{Z}$. This description is also used in [104] and [And]. I prefer to use Grothendieck's initial construction, given in [72], for two reasons:

- It puts categorical issues more clearly in focus.
- It is the model for the later construction of triangulated motives by Voevodsky [135].
- The other construction is ideal for the tensor structure, but less so for the additive structure: it is difficult to describe $(X, p, m) \oplus (Y, q, n)$ when $m \neq n$. Recourse to the projective line formula (6.4.1) is necessary as well as more artificial, cf. [104, 1.14].

Exercise 6.17

(a) Let K/k be an extension. Construct a symmetric monoidal "extension of scalars" functor $i_{K/k} : \mathcal{M}_\sim(k, F) \to \mathcal{M}_\sim(K, F)$ for each of $\sim\ =\ \sim_{\mathrm{rat}}$, $\sim_{\mathrm{alg}}$, $\sim_{\mathrm{tnil}}$ and $\sim\ =\ \sim_{\mathrm{num}}$. (When $\sim\ =\ \sim_H$, one must take "compatible" Weil cohomologies over k and K, cf. [62, 2.2.2].)

(b) If L/K is another extension, define a natural isomorphism $c_{k,K,L}$: $i_{L/k} \xrightarrow{\sim} i_{L/K} \circ i_{K/k}$. How do the $c_{k,K,L}$ behave given three successive extensions (cf. [SGA1, exp. VI, §7] and [62, 1.3])?

(c) In (a), suppose that K/k is finite and separable. Show that the functor from (a) admits a left and right adjoint, sending a motive $h(X)$ to $h(X_{(k)})$ where $X_{(k)}$, for $X \in \mathbf{V}(K)$, is the "naive" restriction of scalars of X [62, th. 4.1].

Exercise 6.18 In Exercise 6.17 (a), suppose that K/k is Galois, with Galois group G.

(a) Give a meaning to the following assertion: the Galois action of G defines a pseudo-action of G on $\mathcal{M}_\sim(K, F)$ (cf. Exercise 6.17 (b)).

(b) A *G-equivariant object of* $\mathcal{M}_\sim(K, F)$ is an object $M \in \mathcal{M}_\sim(K, F)$ equipped with an isomorphism $\alpha_g : M \xrightarrow{\sim} g_* M$ for all $g \in G$, where $\alpha_g, g_* \alpha_h$ and α_{gh} are related by an identity using the isomorphism $c_{g,h} : g_* h_* \xrightarrow{\sim} (gh)_*$ from (a) (specify this identity; cf. [SGA1, exp. VI, §7]). Show that if $M_0 \in \mathcal{M}_\sim(k, F)$, then the object $i_{K/k} M_0$ of $\mathcal{M}_\sim(K, F)$ possesses a natural G-equivariant structure.

(c) Let M be a G-equivariant object of $\mathcal{M}_\sim(K, F)$. Show that there exists an object $M_0 \in \mathcal{M}_\sim(k, F)$, unique up to unique isomorphism, such that $i_{K/k} M_0 \simeq M$ as equivariant G-objects. (Let $\theta : \mathcal{M}_\sim(K, F) \to \mathcal{M}_\sim(k, F)$ be the left or right adjoint of $i_{K/k}$, cf. Exercise 6.17 (c). Define a homomorphism $\rho : G \to \mathrm{Aut}\,\theta(M)$, and then consider the image M_0 of the projector $\frac{1}{|G|} \sum \rho(g)$.)

6.6 Rigidity

Theorem 6.19 *The category* $\mathcal{M}_\sim(k, F)$ *is rigid (Definition A.21).*

Proof Let $\mathcal{M}'$ be the full subcategory of $\mathcal{M}_\sim(k, F)$ whose objects are the $h(X)(n)$ for $X \in \mathbf{V}(k)$ equidimensional and $n \in \mathbf{Z}$. Its Karoubian envelope is $\mathcal{M}_\sim(k, F)$. From Propositions A.22 (3) and A.34, it suffices to show that $h(X)$ is strongly dualisable for X equidimensional.

Let $n = \dim X$. I claim that

$$h(X)^* = h(X)(n)$$

defines a strong dual for X. To see this, we must describe the morphisms η and ε of Definition A.21. Up to a Tate twist, these are morphisms

$$\mathbb{L}^n \xrightarrow{\eta} h(X) \otimes h(X), \quad h(X) \otimes h(X) \xrightarrow{\varepsilon} \mathbb{L}^n.$$

By using the identity $h(X) \otimes h(X) = h(X \times X)$ and Proposition 6.12, these morphisms correspond to classes from

$$A^n_\sim(X \times X, F) = A_n^\sim(X \times X, F).$$

In both cases, we take the diagonal class $[\Delta_X]$. The verification of the identities in Definition A.21 is left as an exercise to the reader. $\qquad\square$

6.7 Jannsen's theorem

Theorem 6.20 (Jannsen [57]) *If F is a field of characteristic zero, then the category $\mathcal{M}_{\mathrm{num}}(k, F)$ is semisimple abelian.*

Before Jannsen's breakthrough, it was known that this statement followed from Grothendieck's standard conjectures. The fact that he could prove it unconditionally was a general surprise. The proof relies on the existence of a Weil cohomology; in its main ideas it is very similar to the proof of Theorem 6.4, which it uses. (See [5] for a version in a more general categorical context.)

We prove Jannsen's theorem. We start with an important particular case which will be needed further on.

Proposition 6.21 *Let K be a field of characteristic zero. For every Weil cohomology H with coefficients in K, the functor $\mathcal{M}_H(k, K) \to \mathcal{M}_{\mathrm{num}}(k, K)$ is essentially surjective, and the category $\mathcal{M}_{\mathrm{num}}(k, K)$ is semisimple abelian.*

Proof Let $X \in \mathbf{V}(k)$ be of pure dimension d. Since $A_H^d(X \times X, K)$ injects into $\mathrm{End}_K(H^*(X))$ via the cycle class homomorphism, it is a finite-dimensional K-algebra; its radical R is therefore nilpotent, and the trace formula shows that $f \sim_{\mathrm{num}} 0$ for all $f \in R$ (cf. [57, th. 1, the proof of c) $\Rightarrow$ b)]). Thus, the surjection $A_H^d(X \times X, K) \to A_{\mathrm{num}}^d(X \times X, K)$ factors through a surjection

$$A_H^d(X \times X, K)/R \to A_{\mathrm{num}}^d(X \times X, K).$$

But the first term is a semisimple K-algebra, thus so is the second term, which suffices to justify the second assertion of the proposition (cf. [57, lemma 2]). For the first assertion, we must show that every idempotent e of the algebra $A_{\mathrm{num}}^d(X \times X, K)$ lifts to an idempotent e of the algebra $A_H^d(X \times X, K)$. But e lifts to an idempotent e_1 of $A_H^d(X \times X, K)/R$ thanks to the structure of semisimple algebras, and e_1 lifts to an idempotent of $A_H^d(X \times X, K)$ thanks to the nilpotence of R (cf. [57, rem. 4]). $\square$

It is now easy to prove Theorem 6.20 in general. Choose a Weil cohomology H whose coefficient field K contains F (cf. Exercise 3.64 (b)). Let $X \in \mathbf{V}(k)$. From Theorem 6.4 (2), we have an isomorphism

$$A_{\mathrm{num}}^d(X \times X, F) \otimes_F K \xrightarrow{\sim} A_{\mathrm{num}}^d(X \times X, K).$$

Since the F-algebra $A_{\mathrm{num}}^d(X \times X, F)$ is finite dimensional (ibid.), its radical R is nilpotent, which implies that $R \otimes_F K$ is contained in the radical of $A_{\mathrm{num}}^d(X \times X, F) \otimes_F K$. But the latter vanishes due to Proposition 6.21, so $R = 0$ and $A_{\mathrm{num}}^d(X \times X, F)$ is semisimple.

Remark 6.22 If H is a *classical* Weil cohomology (§ 3.5.9), the functor $\mathcal{M}_H(k, \mathbf{Q}) \to \mathcal{M}_{\text{num}}(k, \mathbf{Q})$ is already essentially surjective [3, prop. 5]. If k is of characteristic zero, this follows from the comparison theorems by reduction to Betti cohomology, via Proposition 6.21. But in characteristic p this result rests on [26], which is far deeper than Proposition 6.21.

6.8 Specialisation

Let K be a field equipped with a rank 1 discrete valuation v, a valuation ring $\mathcal{O}$, and a residue field k. We would like to define a specialisation functor

$$\mathcal{M}_\sim(K, F) \to \mathcal{M}_\sim(k, F)$$

for at least some adequate pairs $(F, \sim)$. I do not know whether it is possible in this generality, but this is at least possible if we restrict to a full subcategory of $\mathcal{M}_\sim(K, F)$.

Definition 6.23 Write $\mathcal{M}_\sim(K, v, F) \subset \mathcal{M}_\sim(K, F)$ for the thick subcategory generated by the motives of the form $h(X)$, where X is a smooth projective K-variety having good reduction relative to v (cf. Definition 5.3). The objects of $\mathcal{M}_\sim(K, v, F)$ are the *motives with good reduction* (relative to v).

Remarks 6.24

(1) $\mathbf{P}^1$ has good reduction; if X, X' have good reduction, then $X \times X'$ has good reduction. This implies that the subcategory $\mathcal{M}_\sim(K, v, F) \subset \mathcal{M}_\sim(K, F)$ is *rigid*.

(2) Suppose char $k \neq 2$ and let C be the anisotropic projective conic with $aX_0^2 + \pi X_1^2 - X_2^2 = 0$, where $a \in O_K^* - O_K^{*2}$ and π is a uniformiser. Then C does not have good reduction, but $h(C) \simeq h(\mathbf{P}^1)$ in $\mathcal{M}_\sim(K, F)$ whenever 2 is invertible in F. (This example also works in residual characteristic 2, with an appropriate modification.) Thus the motive of a smooth projective K-variety X can have good reduction even if X does not.

Theorem 6.25 *There exists a unique $\otimes$-functor*

$$\text{sp} : \mathcal{M}_\sim(K, v, F) \to \mathcal{M}_\sim(k, F)$$

sending the motive of a smooth projective K-variety X with good reduction to the motive of its special fibre (relative to an arbitrary smooth model) in the following cases:

(1) $\sim$ is rational equivalence, and F is arbitrary.
(2) $\sim$ is algebraic equivalence.
(3) $\sim$ is smash-nilpotence equivalence, and F is arbitrary.
(4) $\sim$ is l-adic homological equivalence where the prime number l is invertible
    in k, and $F \supset \mathbf{Q}$.
(5) char $k = 0$, $\sim$ is a classical Weil cohomology, and $F \supset \mathbf{Q}$.

Proof If the functor exists, its uniqueness and tensoriality are clear.

Case (4) follows from the smooth and proper base change theorem (cf. § 3.6.1), and Case (5) follows from the comparison theorems. Case (3) follows immediately from Case (1). It remains to treat Cases (1) and (2).

I learned the following argument from Yves André. We use specialisation homomorphisms [Ful, ex. 20.3.1]

$$\sigma : A^i_{\sim}(X, F) \to A^i_{\sim}(Y, F) \quad (\sim = \sim_{\mathrm{rat}}, \sim_{\mathrm{alg}})$$

where X/K has good reduction and Y is the special fibre of a smooth model $\mathcal{X}$ of X.

To be more precise, in loc. cit., Fulton only treats the case of rational equivalence. But its construction rests on the exact sequences [Ful, prop. 1.8]

$$A^{n-1}_{\sim}(Y, F) \xrightarrow{i_*} A^n_{\sim}(\mathcal{X}, F) \xrightarrow{j^*} A^n_{\sim}(X, F) \to 0 \qquad (6.8.1)$$

(where $j : X \to \mathcal{X}, i : Y \to \mathcal{X}$ are closed or open immersions, corresponding to our situation), and the identity

$$i^* i_* = 0 \qquad (6.8.2)$$

deduced from [Ful, cor. 6.3]. But the exact sequence (6.10.1) holds just as well for algebraic equivalence [Ful, ex. 10.3.4][3], and clearly so does the identity (6.8.2).

If X' is another variety with good reduction, equipped with a smooth model $\mathcal{X}'$ with special fibre Y', then $\mathcal{X} \otimes_{\mathcal{O}} \mathcal{X}'$ is a smooth model of $X \times X'$, with special fibre $Y \times Y'$, and σ induces a homomorphism

$$\mathrm{sp} : \mathrm{Corr}_{\sim}(K, F)(X, X') \to \mathrm{Corr}_{\sim}(k, F)(Y, Y').$$

If X'' is a third variety with good reduction, equipped with a smooth model $\mathcal{X}''$ with special fibre Y'', and if

$$\alpha \in \mathrm{Corr}_{\sim}(K, F)(X, X'), \quad \beta \in \mathrm{Corr}_{\sim}(K, F)(X', X'')$$

---

[3] This argument is insufficient in mixed characteristic, where one needs to lift smooth curves and their rational points from k to $\mathcal{O}$ (curves are sufficient by [58, p. 89, Cor. 6.11 (2) (3)]). For this, one may use [80, Th. 1].

are correspondences, then the compatibilities proved in [Ful, 20.3] imply that

$$sp(\beta \circ \alpha) = sp(\beta) \circ sp(\alpha).$$

In particular, take $X = X' = X''$, $\alpha = \beta = \Delta_X$, and $\mathcal{X}'' = \mathcal{X}$: we get that $sp(\alpha)$ and $sp(\beta)$ are mutually inverse isomorphisms from $h(Y)$ to $h(Y')$. By letting the smooth models of X vary, we get a *transitive system of isomorphisms* between the motives $h(Y)$, where Y runs through the corresponding special fibres. This shows that the motive $h(Y)$ is well-determined up to unique isomorphism, and that sp defines a functor; this functor is clearly monoidal, reusing the compatibility of σ with the product [Ful, 20.3].

The construction of sp now automatically extends to effective motives, and then to all motives. □

Remark 6.26 The case of numerical equivalence is not clear; it would follow from the standard conjectures. See [3, 3] for a way to get around this problem.

Exercise 6.27

(a) Let L/K be a finite Galois extension, and let w be an extension of v to L. Write l for the residue field of w. Show that the functor sp : $\mathcal{M}_\sim(L, w, F) \to \mathcal{M}_\sim(l, F)$ is equivariant relative to the pseudo-action of the decomposition group of w on $\mathcal{M}_\sim(L, w, F)$ and to the pseudo-action of $\mathrm{Gal}(l/k)$ on $\mathcal{M}_\sim(l, F)$, cf. Exercise 6.18 (a). (First, give a meaning to the statement.)

(b) We say that a smooth projective K-variety X has *potentially good reduction* if there is a finite separable extension L/K and an extension w of v to L such that X_L has good reduction (relative to w). Extend Theorem 6.25 to the thick subcategory $\mathcal{M}_\sim(K, v, F)^{\mathrm{pot}}$ of $\mathcal{M}_\sim(K, F)$ generated by the motives of varieties with potentially good reduction. (Use Exercise 6.18 (c).)

6.9 Motivic theory of weights (pure case)

Suppose that F is a field of characteristic zero, and suppose we are given a Weil cohomology H^* with coefficients in $K \supset F$.

Definition 6.28 Let $X \in \mathbf{V}(k)$ and let $i \in \mathbf{Z}$. The *i-th Künneth projector of X* (relative to H) is the projection p^i of

$$H^*(X) = \bigoplus_j H^j(X)$$

onto $H^i(X)$. This is an idempotent endomorphism of $H^*(X)$.

Note that this definition extends to every motive $M \in \mathcal{M}_H(k, F)$: we may talk of *Künneth projectors of M*.

Definition 6.29 Let $M \in \mathcal{M}_H(k, F)$, and let $i \in \mathbf{Z}$.

a) We say that M is *pure of weight i* if $H^j(M) = 0$ for $j \neq i$. We abbreviate this by $w(M) = i$.

b) We say that *the i-th Künneth projector of M is algebraic* if we have

$$p^i = H(p^i_{alg}) \quad \text{for } p^i_{alg} \in \mathrm{End}(M).$$

We abbreviate this as: "p^i_M is algebraic". If $M = h(X)$ for $X \in \mathbf{V}(k)$, then we write p^i_X for p^i_M.

If this condition is satisfied, then the representative p^i_{alg} is unique; we may then write p^i_M without ambiguity. We also write $M^{(i)} = \mathrm{Im}\, p^i_M$: this is the *part of weight i* of M. If $M = h(X)$, we write $h^i(X)$ for $M^{(i)}$.

Grothendieck's standard conjectures imply that the i-th Künneth projector of M is algebraic for all M and all i. Let us indicate what is known unconditionally, i.e. without assuming these conjectures:

Lemma 6.30 *Let $M, N \in \mathcal{M}_H(k, F)$ and let $i, j \in \mathbf{Z}$.*

(1) *If p^i_M and p^i_N are algebraic, then $p^i_{M \oplus N}$ is algebraic.*

(2) *If p^i_M is algebraic, then $p^i_{M'}$ is algebraic for every direct summand M' of M.*

(3) *If the p^*_M and the p^*_N are algebraic, then the $p^*_{M \otimes N}$ are algebraic.*

(4) *If p^i_M is algebraic, then $p^{-i}_{M^*}$ is algebraic.*

Proof (1), (3) and (4) follow from the identities (in $\mathbf{Vec}^*_K$)

$$p^i_{M \oplus N} = p^i_M \oplus p^i_N, \quad p^k_{M \otimes N} = \sum_{i+j=k} p^i_M \otimes p^j_N, \quad p^{-i}_{M^*} = {}^t p^i_M.$$

For (2), let $p \in \mathrm{End}(M)$ be a projector with image M'. We note that p^i_M is *central* in $\mathrm{End}(H^*(M))$, and thus *a fortiori* in the subring $\mathrm{End}(M)$. So p^i_M commutes with p and pp^i_M is still a projector; we see immediately that it is $p^i_{M'}$. $\qquad \square$

Theorem 6.31 *Let H be a Weil cohomology satisfying the axiom of Remark 3.43 (for example, a classical Weil cohomology, § 3.5.9).*

(1) *For all $X \in \mathbf{V}(k)$, of pure dimension n, p_X^i is algebraic for $i = 0, 1, 2n - 1, 2n$. If X is geometrically connected, then $h^{2n}(X) \simeq \mathbb{L}^n$.*

(2) *If $n = 2$, all the p_X^i are algebraic.*

(3) *If A is an abelian variety, all the p_A^i are algebraic.*

Proof (1) Write k_X for the field of constants of X, that is to say, the algebraic closure of k in $k(X)$. Since X is smooth, k_X/k is separable, so $h(\mathrm{Spec}\, k_X)$ is defined. I claim

$$h_X^0 = h(k_X)$$
$$= 1 \quad \text{if } k_X = k.$$

First note that $H^0(k_X) \xrightarrow{\sim} H^0(X)$ (axiom (v) of a Weil cohomology). Now consider a closed point $i_x : x \hookrightarrow X$ with residue field E. Let $f : \mathrm{Spec}\, E \to \mathrm{Spec}\, k_X$ be the canonical morphism. Then the composition

$$h(k_X) \xrightarrow{p'^*} h(X) \xrightarrow{i_x^*} h(x) \xrightarrow{f_*} h(k_X)$$

is equal to $\deg f \cdot 1_{h(k_X)}$ thanks to the formula of Proposition 2.22 (which holds in full generality). Then $\frac{1}{\deg f} p'^* f_* i_x^*$ is the desired projector p_X^0. By duality, we obtain that p_X^{2n} is algebraic and the value of $h^{2n}(X)$.

The cases of p_X^1 and p_X^{2n-1} are much more delicate; they are due to Grothendieck and generalised by Jacob Murre to rational equivalence (the two proofs rely on the same theorem of Weil). I refer the reader to [104, § 4] for the details.

(2) follows from (1) since

$$p_X^2 = 1_{h(X)} - p_X^0 - p_X^1 - p_X^3 - p_X^4.$$

(3) This theorem is due to Lieberman–Kleiman, and was generalised by Shermenev, Deninger–Murre and Künnemann to rational equivalence: I refer to [32] and [74] for the details. $\qquad\square$

Corollary 6.32 *For every $M \in \mathcal{M}_H^{\mathrm{eff}}$, p_M^i is algebraic for $i = 0, 1$.*

Proof By Lemma 6.30 (2), we reduce to $M = h(X)$ for $X \in \mathbf{V}(k)$ and apply Theorem 6.31 (1). $\qquad\square$

Finally, the following theorem is much deeper: it relies on Deligne's proof of the Riemann hypothesis [30].

Theorem 6.33 (Katz–Messing [69]) *If k is finite, the Künneth projectors are algebraic for all $X \in \mathbf{V}(k)$ and $H = H_l$ for all $l \neq \mathrm{char}\, k$, as well as for* *crystalline cohomology. Furthermore, p_X^i is given by an algebraic cycle independent of the chosen Weil cohomology, and which is a linear combination of powers of Frobenius.*

6.10 Example: Artin motives

Definition 6.34 An *Artin motive* is a motive isomorphic to a direct summand of $h(X)$, where X is of dimension 0.

A smooth k-scheme of dimension 0 is of the form Spec E, where E is an étale k-algebra, i.e. a product $\prod_{i=1}^{r} E_i$ of finite separable extensions of k.

Proposition 6.35 *Let $\mathcal{M}_{\sim}^0(k, F) \subset \mathcal{M}_{\sim}(k, F)$ be the full subcategory consisting of the Artin motives. Then:*

(1) *$\mathcal{M}_{\sim}^0(k, F) \subset \mathcal{M}_{\sim}^{\mathrm{eff}}(k, F)$.*
(2) *The category $\mathcal{M}_{\sim}^0(k, F)$ is a rigid symmetric monoidal subcategory of $\mathcal{M}_{\sim}(k, F)$.*
(3) *$\mathcal{M}_{\sim}^0(k, F)$ does not depend on the choice of the adequate equivalence $\sim$, in the following sense: if $\sim \geq \sim'$, then the functor $\mathcal{M}_{\sim}^0(k, F) \to \mathcal{M}_{\sim'}^0(k, F)$ is an equivalence of categories.*
(4) *For $\sim = \sim_H$, where H is a Weil cohomology satisfying the axioms of Remark 3.43, we have*

$$\mathcal{M}_H^0(k, F) = \{M \in \mathcal{M}_H^{\mathrm{eff}} \mid w(M) = 0\}$$

(cf. Definition 6.29 a)).

Proof (1) and (2) are clear; for (3) we reduce to proving

$$\mathcal{M}_{\sim}(h(X), h(Y)) \xrightarrow{\sim} \mathcal{M}_{\sim'}(h(X), h(Y))$$

if dim $X = $ dim $Y = 0$, which is clear since these are just cycles of dimension 0. In (4), the inclusion $\subset$ is clear; conversely, if M is effective and $w(M) = 0$, we write it as a direct summand of $h(X)$ for X smooth projective. Therefore M is a direct summand of $h^0(X)$ (cf. Theorem 6.31 (1)). $\qquad\square$

Theorem 6.36 *Set $G = \mathrm{Gal}(k_s/k)$. The category $\mathcal{M}_{\sim}^0(k, F)$ is equivalent to the (symmetric monoidal) category $\mathrm{Rep}_F(G)$ of continuous linear representations of G with coefficients in F.*

Proof If dim $X = 0$, Galois theory (as reformulated by Grothendieck) associates to X the finite set $X(k_s)$ equipped with a (continuous) G-action, giving the "permutation" representation

$$\rho_X = F[X(k_s)] \in \operatorname{Rep}_F(G).$$

As $\operatorname{Rep}_F(G)$ is pseudo-abelian, this extends to a functor

$$\rho : \mathcal{M}^0_\sim(k, F) \to \operatorname{Rep}_F(G)$$

which is clearly symmetric monoidal. It remains to see that ρ is an equivalence of categories.

To see full faithfulness, we reduce by rigidity (Proposition 6.35 (2)) to checking that

$$\mathcal{M}_\sim(\mathbf{1}, h(X)) \xrightarrow{\sim} \operatorname{Rep}_F(G)(F, \rho_X) \tag{6.10.1}$$

if $X = \operatorname{Spec} E$, where E is a finite separable extension of k. The group on the left is $A^0_\sim(X, F) \simeq F$ and the one on the right is

$$\operatorname{Hom}_G(F, F[X(k_s)]) \simeq F$$

since the action of G on $X(k_s) = \operatorname{Hom}_k(E, k_s)$ is transitive. We see right away that (6.10.1) sends the generator $[X]$ of the term on the left to the element $\sum_{x \in X(k_s)} x$ of the term on the right.

It remains to show essential surjectivity. Let $\rho \in \operatorname{Rep}_F(G)$. By continuity, ρ factors through $\bar{G} = \operatorname{Gal}(E/k)$, where E is a finite Galois extension of k. Then ρ is a direct summand of the regular representation $r_{\bar{G}} = \rho_{\operatorname{Spec} E}$. $\qquad\square$

6.11 Example: h^1 of abelian varieties

Let A be an abelian variety over k. By Theorem 6.31 (3), the Künneth projectors of $h(A)$ are algebraic; by Deninger–Murre [32], we even get Chow–Künneth projectors p^i_A modulo rational equivalence, which are natural for homomorphisms of abelian varieties. This defines a (contravariant) functor

$$h^1 : \operatorname{Ab}(k)^{\operatorname{op}} \to \mathcal{M}_{\operatorname{rat}}(k, \mathbf{Q})$$

where $\operatorname{Ab}(k)$ is the category of abelian k-varieties (objects: abelian varieties; morphisms: homomorphisms of abelian varieties).

Theorem 6.37 *The functor h^1 is additive. It induces a fully faithful functor*

$$h^1 : \operatorname{Ab}^0(k)^{\operatorname{op}} \to \mathcal{M}_{\operatorname{rat}}(k, \mathbf{Q})$$

where $Ab^0(k)$ *has the same objects as* $Ab(k)$, *the morphisms being tensored with* **Q**. *Its composition with the projection* $\mathcal{M}_{rat}(k, \mathbf{Q}) \to \mathcal{M}_{\sim}(k, \mathbf{Q})$ *is still fully faithful for any adequate equivalence relation* $\sim$. *If H is a Weil cohomology satisfying the axiom of Remark 3.43, then the essential image of h^1 in* $\mathcal{M}_H^{eff}$ *is formed of the motives of pure weight 1.*

Proof By the construction of the projectors p_A^i, we have

$$p_A^i \circ n_A = n^i \, p_A^i$$

for all $i \geq 0$ and $n \in \mathbf{Z}$, where n_A is the endomorphism on A given by multiplication by n [32, th. 3.1]. We deduce easily that h^1 is additive; hence it extends to $Ab^0(k)$. Since $\mathcal{M}_{rat}(k, \mathbf{Q}) \to \mathcal{M}_{\sim}(k, \mathbf{Q})$ is full for all $\sim$, and since $\sim_{num}$ is the coarsest adequate equivalence, it suffices to show that h^1 is full for $\sim = \sim_{rat}$ and faithful for $\sim = \sim_{num}$. For this, we may reduce to the case of two Jacobians of curves J and J' by Theorem 3.27, and the two statements follow from the Weil isomorphism (Theorem 3.38). Finally, the last assertion is proved as in Proposition 6.35 (4). $\qquad\square$

Exercise 6.38

(a) Show that in Theorem 6.37, a quasi-inverse of h^1 is induced by $X \mapsto$ $Alb(X)$ (cf. § 3.1.5).

(b) Show that for all $X \in \mathbf{V}(k)$, there is a canonical isomorphism

$$h^1(X) \simeq h^1(Alb(X))$$

which is induced by the Albanese morphism $X \to Alb(X)$ given by a rational point $x \in X(k)$, whenever such a point exists (and the isomorphism is then independent of the choice of x).

6.12 The zeta function of an endomorphism

6.12.1 The case of a sufficiently fine equivalence relation

By applying Theorem A.41, we get

Theorem 6.39 *Suppose F is a field of characteristic zero. Let $(F, \sim)$ be an adequate pair such that $\sim \geq \sim_H$ for a Weil cohomology H with coefficients in a field extension K of F. For all $M \in \mathcal{M}_{\sim}(k, F)$ and every endomorphism $f \in End(M)$, the function $Z(f, t)$ from Definition A.40 is a rational function with coefficients in F. If f is an isomorphism, we have a functional equation of the form*

$$Z({}^t f^{-1}, t^{-1}) = \det(f)(-t)^{\chi(M)} Z(f, t)$$

where ${}^t f \in \text{End}(M^*)$ is the transpose of f and $\det(f)$ is a scalar, given by the formula of Theorem A.41.

Proof The Weil cohomology H provides a symmetric monoidal functor $H^* = \mathcal{M}_\sim(k, F) \to \mathbf{Vec}_K^*$ for a field K containing F. $\square$

6.12.2 The case of numerical equivalence

Grothendieck's main standard conjecture [44] predicts that homological equivalence coincides with numerical equivalence for every Weil cohomology. Since this conjecture is unresolved, the extension of Theorem 6.39 to motives modulo numerical equivalence is delicate. Nonetheless, we have

Theorem 6.40 *For every field F of characteristic zero, Theorem 6.39 remains true if we let $\sim$ be numerical equivalence.*

Proof We proceed as in the proof of Theorem 6.20: choose a Weil cohomology H with coefficients in a field extension K of F (Exercise 3.64 (b)). Let M_K be the image of $M \in \mathcal{M}_{\text{num}}(k, F)$ in $\mathcal{M}_{\text{num}}(k, K)$. By Proposition 6.21, we may lift M_K to a motive of $\mathcal{M}_H(k, K)$, which proves Theorem 6.40, except that *a priori* the rational function considered has coefficients in K; we conclude as usual by using Exercise 3.71. $\square$

We now make the link with the theory of weights. We abbreviate the notation $\mathcal{M}_\sim(k, \mathbf{Q})$ to $\mathcal{M}_\sim$.

Proposition 6.41 ([3, prop. 6]) *Let H be a classical Weil cohomology. Write $\mathcal{M}_H^*$ for the full subcategory of $\mathcal{M}_H$ formed from those M all of whose Künneth projectors are algebraic, and let $\mathcal{M}_{\text{num}}^*$ be its essential image in $\mathcal{M}_{\text{num}}$. Then*

(1) *$\mathcal{M}_{\text{num}}^*$ does not depend on the choice of H.*
(2) *For $M \in \mathcal{M}_{\text{num}}^*$ arising from $\tilde{M}_H \in \mathcal{M}_H^*$ (cf. Remark 6.22), the image p_M^i of $p_{\tilde{M}_H}^i$ in $\text{End}(M)$ does not depend on the choice of H nor on the choice of $\tilde{M}_H$.*
(3) *If k is finite, then $\mathcal{M}_{\text{num}}^* = \mathcal{M}_{\text{num}}$.*

This statement is incompletely proved in [3]; the proof (following O. Gabber) is completed in [4, app. B]. Note that (3) simply repeats Theorem 6.33.

Corollary 6.42 *For every simple motive* $S \in \mathcal{M}^*_{\text{num}}$*, there is a unique* $i \in \mathbf{Z}$ *such that* $p^i_S \neq 0$*: this is the* weight *of S.*   □

Definition 6.43 A motive $M \in \mathcal{M}^*_{\text{num}}$ is *pure of weight i* if it is a direct sum of simple motives of weight i.

Corollary 6.44 *Let* $M \in \mathcal{M}^*_{\text{num}}$ *be pure of weight i. Then for every endomorphism f of M, we have* $Z(f, t) = P(t)^{(-1)^{i+1}}$*, where* $P \in F[t]$*. We also have the functional equation from Theorem* 6.39.

Proof This follows immediately from the trace formula (Lemma A.37).   □

Exercise 6.45

(a) Let C, C' be smooth projective curves over k. Construct an isomorphism

$$\mathcal{M}_{\sim}(h^1(C), h^1(C')) \simeq \mathrm{Corr}_{\equiv}(C', C) \otimes \mathbf{Q},$$

where $\mathrm{Corr}_{\equiv}(C', C)$ is the group of correspondence classes of Definition 2.28.

(b) Suppose that $C = C'$. Show that for all $f \in \mathrm{End}\, h^1(C)$, we have $\mathrm{Tr}(f) = \sigma(f)$, where σ is the trace of Definition 2.27 and Tr is the categorical trace of Definition A.35.

6.13 The case of a finite base field

Suppose now that $k = \mathbf{F}_q$. For every $X \in \mathbf{V}(k)$, write π_X for the Frobenius endomorphism viewed as an algebraic correspondence (the transpose of the graph of the absolute Frobenius of X). Then π_X is *central* in $\mathrm{Corr}^0_{\text{rat}}(X, X)$; more generally, for every correspondence $\gamma \in \mathrm{Corr}^0_{\text{rat}}(X, Y)$, the diagram

$$
\begin{array}{ccc}
h(X) & \xrightarrow{\gamma} & h(Y) \\
\pi_X \downarrow & & \pi_Y \downarrow \\
h(X) & \xrightarrow{\gamma} & h(Y)
\end{array}
$$

is commutative. Moreover, $\pi_{X \times Y} = \pi_X \otimes \pi_Y$. It follows that $X \mapsto \pi_X$ extends to a $\otimes$-*automorphism of the identity functor*

$$\mathcal{M}_{\text{rat}}(k, \mathbf{Q}) \ni M \mapsto \pi_M \in \mathrm{Aut}(M)$$

satisfying the identities

$$\pi_{M \oplus N} = \pi_M \oplus \pi_N, \quad \pi_{M \otimes N} = \pi_M \otimes \pi_N, \quad \pi_{M^*} = {}^t\pi_M^{-1}. \qquad (6.13.1)$$

(The last identity is a general property of a $\otimes$-automorphism of the identity functor, cf. [Saa, I, (3.2.3.6)].)

Recall also that $\mathcal{M}_{\text{num}} = \mathcal{M}^*_{\text{num}}$ by Proposition 6.41 (3).

Proposition 6.46

(1) *The endomorphism $\pi_{\mathbb{L}}$ of the Lefschetz motive $\mathbb{L}$ is multiplication by q.*

(2) *If X is geometrically connected of dimension n, the endomorphism $\pi_{h^{2n}(X)}$ is multiplication by q^n.*

(3) *If $M \in \mathcal{M}^{\text{eff}}_{\text{num}}(k, \mathbf{Q})$ is pure of weight i (Definition 6.43), we have*

$$\det(\pi_M) = \pm q^{i\chi(M)/2},$$

where $\det(\pi_M)$ is the rational number appearing in the functional equation of Theorem 6.39.

Proof (1) Let $p = p^2_{\mathbf{P}^1}$ be the second Künneth projector of the projective line: it cuts out $\mathbb{L}$ on the motive of $\mathbf{P}^1$. We must then check that

$$\pi_{\mathbf{P}^1} p = q p.$$

But p is the class of $0 \times \mathbf{P}^1$ in $CH^1(\mathbf{P}^1 \times \mathbf{P}^1)$, the transpose of the graph of the composition $\mathbf{P}^1 \to \{0\} \hookrightarrow \mathbf{P}^1$. If (t_1, t_2) are coordinates of $\mathbf{P}^1 \times \mathbf{P}^1$, the equation of this divisor is $t_1 = 0$. When we apply $\pi_{\mathbf{P}^1}$ we get the equation $t_1^q = 0$, hence the conclusion. (2) now follows from a), Theorem 6.31 and (6.13.1). Finally, for (3) we use the formula of Theorem A.41, the Riemann hypothesis for $H^i(X)$ and the fact that the inverse roots of the characteristic polynomial of π_M form a (Galois-invariant!) subset of the inverse roots of the characteristic polynomial of $\pi_{h^i(X)}$. $\square$

Remarks 6.47

(1) In particular, if i is odd, then $\chi(M)$ is even in Proposition 6.46 (3), at least if q is not a square. (This is conjectured to be true also if q is a square: see Exercise 6.56 (c).)

(2) If M is of the form $h^i(X)$ with i odd, the sign in the expression of $\det(\pi_M)$ is $+1$. In fact, for l a prime not dividing q, Poincaré duality and the hard Lefschetz theorem equip $H^i_l(X)$ with a Galois-equivariant perfect pairing:

$$H^i_l(X) \times H^i_l(X) \to \mathbf{Q}_l(-i)$$

which is *alternating* since i is odd (Corollary 5.42); in particular, $d = \dim H^i_l(X)$ is even. The determinant of a symplectic matrix being equal to 1, that of π_M is equal to $(\sqrt{q^i})^d$. See Remark 3.66 for the case of even i.

(3) One may consult [60, III] for a different approach to Proposition 6.46 (3) (and a generalisation to "motivic zeta functions"), using the notion of the determinant of a motive.

Definition 6.48 For $M \in \mathcal{M}_{\text{num}}(k, \mathbf{Q})$, we write

$$Z(M, t) = Z(\pi_M, t).$$

We also set $\zeta(M, s) = Z(M, q^{-s})$: this is the *zeta function* of M.

Lemma 6.49

(1) *If $M = h(X)$ for $X \in \mathbf{V}(k)$, we have $Z(M, t) = Z(X, t)$.*
(2) *For all $M \in \mathcal{M}_{\text{num}}(k, \mathbf{Q})$, we have*

$$Z(M(1), t) = Z(M, q^{-1}t), \quad \zeta(M(1), s) = \zeta(M, s + 1).$$

Proof (1) is clear, cf. the proof of Theorem 3.65; (2) is an easy computation. $\square$

Theorem 6.50 *Suppose $M \in \mathcal{M}_{\text{num}}^{\text{eff}}(k, \mathbf{Q})$ is pure of weight i. Then $Z(M, t) = P(t)^{(-1)^{i+1}}$ where $P(0) = 1$ and $P \in \mathbf{Z}[t]$. We have a functional equation of the form*

$$Z(M^*, t^{-1}) = \det(\pi_M)(-t)^{\chi(M)} Z(M, t), \quad \det(\pi_M) = \pm q^{i\chi(M)/2},$$

where $\chi(M)$ is odd and the sign of $\det(\pi_M)$ is $+$ if i is odd and M is of the form $h^i(X)$, cf. Remarks 6.47 (1) and (2). Furthermore, the inverse roots of P are Weil numbers of weight i.

Proof Corollary 6.44 gives the first and second part of the statement, using (6.13.1) for the functional equation.

For the last assertion (including $P \in \mathbf{Z}[t]$), we write M as a direct summand of $h(X)$ for $X \in \mathbf{V}(k)$ (not necessarily connected); then M is even a direct summand of $h^i(X)$, and P divides the polynomial P_i associated to X. The conclusion then follows from the Riemann hypothesis. $\square$

In particular, for smooth projective X,

$$Z(h^i(X), t) = P_i(t)^{(-1)^{i+1}} \tag{6.13.2}$$

where P_i is the ith polynomial in the factorisation of $Z(X, t)$ (§ 3.3).

Remarks 6.51

(1) The statement still holds if we remove the condition that M be effective, but P may then take coefficients in $\mathbf{Z}[1/q]$: thus, for example, $Z(\mathbb{T}, t) = (1 - q^{-1}t)^{-1}$.
(2) Theorem 6.50 extends to the case where the coefficients $\mathbf{Q}$ are replaced by a number field F; the polynomial P then takes coefficients in O_F if M is effective, and in $O_F[1/q]$ in general.
(3) For $i = 0, 1$, Corollary 6.44 and Theorem 6.50 do not depend on the results of Deligne and Katz–Messing, but "only" on those of Weil (the formula for morphisms between Jacobians of curves and the Riemann hypothesis for curves): cf. Theorems 6.36 and 6.37.

6.14 The Tate conjecture

We continue to take $k = \mathbf{F}_q$.

Conjecture 6.52 (Tate) *For every smooth projective k-variety X and every integer $i \geq 0$, we have*

$$\mathrm{ord}_{s=i}\, \zeta(X, s) = -\,\mathrm{rg}\, A^i_{\mathrm{num}}(X).$$

(Recall from theorem 6.4 that the group $A^i_{\mathrm{num}}(X)$ is free and of finite type.)

This conjecture holds for $i = 0$ due to the "Riemann hypothesis" part of the Weil conjectures. For more details on the case $i > 0$, I refer to Tate's exposition [130]. In particular, for a prime number $l \nmid q$, consider the following statements, where $G = \mathrm{Gal}(\bar{k}/k)$:

$T^i(X, l)$ The homomorphism induced from the cycle class

$$CH^i(X) \otimes \mathbf{Q}_l \to H^{2i}_l(X)(i)^G,$$

is surjective.
$S^i(X, l)$ The composition

$$H^{2i}_l(X)(i)^G \hookrightarrow H^{2i}_l(X)(i) \to H^{2i}_l(X)(i)_G$$

is bijective.
$SS^i(X, l)$ The action of G on $H^i_l(X)$ is semisimple.

We clearly have $SS^{2i}(X, l) \Rightarrow S^i(X, l)$, and

Theorem 6.53 ([130, th. 2.9]) *For all l, Conjecture 6.52 for (X, i) is equivalent to $T^i(X, l) + T^{d-i}(X, l) + S^i(X, l)$; this implies the standard conjecture "homological equivalence = numerical equivalence" in codimensions i and $d - i$. Moreover, $S^i(X, l) \iff S^{d-i}(X, l)$.*

Theorem 6.54 ([59, th. 6]) *We have the equivalence*

$$S^d(X \times X, l) \iff SS^i(X, l) \text{ for all } i.$$

The statements $T^1(X, l)$ and $SS^i(X, l)$ are known when X is an abelian variety [127]: this is a variant of Theorem 3.8. Consequently, Conjecture 6.52 is known for $i = 1$ when X is an abelian variety.

Exercise 6.55 In this exercise, we consider the enriched realisation functor

$$R_l^* : \mathcal{M}_l(k, \mathbf{Q}_l) \to \operatorname{Rep}^*(G, \mathbf{Q}_l)$$

induced by the Weil cohomology H_l with values in the abelian category of $\mathbf{Q}_l$-adic $\mathbf{Z}$-graded representations of G.

(a) (cf. [And, prop. 7.3.1.3 1)]) Show that Conjecture $T^i(X, l)$ for all (X, i) is equivalent to the following statement:
T(M,N,l) The homomorphism

$$\mathcal{M}_l(k, \mathbf{Q}_l)(M, N) \to \operatorname{Rep}^*(G, \mathbf{Q}_l)(R_l^*(M), R_l^*(N))$$

is surjective for all $M, N \in \mathcal{M}_l(k, \mathbf{Q}_l)$.
In other words, R_l^* is *full*. (Prove the equivalence $T(M, N, l) \iff T(\mathbf{1}, M^* \otimes N, l)$, and then reduce to $T^*(X, l)$ for suitable X.)

(b) Show similarly that Conjecture $SS^i(X, l)$ for all (X, i) implies that
SS(M,l) $R_l^*(M)$ is semisimple for all $M \in \mathcal{M}_l(k, \mathbf{Q}_l)$.

(c) Let $M, N \in \mathcal{M}_l(k, \mathbf{Q}_l)$. Suppose that $\dim R_l^i(M) = \dim R_l^i(N)$ for all i and that the eigenvalues of the action of the Frobenius on these vector spaces coincide, with multiplicities. Given Conjectures $T(M, N, l)$, $SS(M, l)$, and $SS(N, l)$, deduce that $M \simeq N$.

(d) Show that Conjecture 6.52 implies the following statement: for every $M \in \mathcal{M}_{\text{num}}(k, \mathbf{Q})$ and $i \in \mathbf{Z}$, we have

$$\operatorname{ord}_{s=i} \zeta(M, s) = -\dim_{\mathbf{Q}} \mathcal{M}_{\text{num}}(M, \mathbb{L}^i).$$

(Reuse the proof of [130, th. 2.9].)

Exercise 6.56 In this exercise, we examine what happens to the sign of the functional equation in Theorem 6.50 when we replace $h^i(X)$ by an arbitrary motive of odd weight i.

(a) Let $M \in \mathcal{M}_{\text{num}}^{\text{eff}}(k, \mathbf{Q})$ be pure of weight i. Show that one can write $M \simeq M' \oplus M''$, where the eigenvalues of $\pi_{M'}$ (resp. of $\pi_{M''}$) are of the form $\pm q^{i/2}$ (resp. are not of this form). (Construct the corresponding idempotent as a polynomial in π_M.)

(b) If $M' = 0$, show that $\chi(M)$ is even, that $\det(\pi_M) > 0$, and that the sign of the functional equation of $\zeta(M, s)$ is $+1$. (Use the formulae from Theorem 6.39 and Theorem A.41, as well as Remark 5.32.)

(c) Suppose that $M'' = 0$ and i is odd. Let $\varepsilon \in \{+, -\}$. Following Deuring [35] (see also Honda [53]), there exists a supersingular elliptic curve E_ε defined over k such that π_{E_ε} has eigenvalue $\varepsilon q^{i/2}$. Using Exercise 6.55 (c), show that if we assume Conjecture 6.52, then M is a sum of motives of the form $h^1(E_\varepsilon)(\frac{1-i}{2})$. Deduce that the conclusions of (b) remain true.[4]

(d) What happens when i is even?

6.15 Coronidis loco

Let us derive from the above a motivic proof of Weil's theorem 4.72 (Artin's conjecture over function fields): in view of Remark 6.51 (3), it only uses Weil's results.

Let K be a function field in one variable over $\mathbf{F}_q$: up to enlarging $\mathbf{F}_q$, we may suppose it to be algebraically closed in K. Let L be a finite regular extension of K that is Galois with Galois group G. Let $\rho : G \to GL_n(\mathbf{C})$ be a nontrivial linear irreducible representation. Then ρ is defined over a number field F. By choosing a finite place λ of F over a prime number l which does not divide q, we may view ρ as a representation taking values in the completion F_λ, a finite extension of $\mathbf{Q}_l$.

Let $p : \tilde{C} \to C$ be the (ramified) covering, Galois with Galois group G, of the smooth projective models of K and L. The group G operates on the motive $h(\tilde{C}) \in \mathcal{M}_{\text{num}}(\mathbf{F}_q, F_\lambda)$, and ρ cuts out a motive M_ρ; the trace formula (extended to F_λ coefficients) shows that

$$L(\rho^\vee, t) = Z(M_\rho, t).^5$$

By hypothesis on ρ, M_ρ is a direct summand of the image of the projector $\pi = 1 - \frac{1}{|G|} p^* p_*$. Write

---

[4] By examining the proof, one sees that Conjecture 6.52 is in fact only used in codimension $\frac{i+1}{2}$.

[5] Note that Artin L-functions are defined in terms of arithmetic Frobeniuses, contrary to L-functions of l-adic sheaves and of motives: to pass from one convention to the other, we must replace ρ by its dual.

$$h(C) = \mathbf{1} \oplus h^1(C) \oplus \mathbb{L},$$
$$h(\tilde{C}) = \mathbf{1} \oplus h^1(\tilde{C}) \oplus \mathbb{L}$$

(we used the hypothesis that the extension L/K is regular). Since $\mathrm{Im}(1 - \pi) = h(C)$, we see that M_ρ is a *direct summand of* $h^1(\tilde{C})$; so $Z(M_\rho, t)$ is a polynomial dividing $P_1(\tilde{C})$ (and even $P_1(\tilde{C})/P_1(C)$). Thus $\zeta(\rho^\vee, s) = Z(M_\rho, q^{-s})$ is an entire function whose zeros lie on the line $\Re(s) = 1/2$.

Exercise 6.57 Suppose that the extension L/K is no longer necessarily regular. Let k be the algebraic closure of $\mathbf{F}_q$ in L. Show that Theorem 4.72 remains true, provided the representation ρ does not factor through $\mathrm{Gal}(kK/K)$.

Appendix A
Karoubian and monoidal categories

A.1 Karoubian categories

A.1.1 Idempotent endomorphisms (or projectors)

Definition A.1 Let $\mathcal{C}$ be a category, and let C be an object of $\mathcal{C}$. An endomorphism p of C is *idempotent* if $p^2 = p$. We also say that p is a *projector*.

Definition A.2 Let $\mathcal{C}, C, p$ be as above. If it exists, the *image* of p, denoted by Im p, is the kernel of the pair of parallel arrows $(1_C, p)$.

(This definition is justified by the following calculation, itself justified by the Yoneda lemma: if $x \in C$ is of the form $p(y)$, then $p(x) = p^2(y) = p(y) = x$.)

Lemma A.3 *Suppose that p has an image. Then* Im p *is canonically a retract of C.*

Proof By definition, we have a canonical morphism Im $p \to C$ equalising the two arrows $1_X, p$. Conversely, the endomorphism $p : C \to C$ equalises these two arrows, and thus factors canonically through Im p. We immediately see that the composition

$$\text{Im } p \to C \to \text{Im } p$$

is equal to the identity. $\square$

Definition A.4 The category $\mathcal{C}$ is called *Karoubian* if every projector has an image.

A.1.2 Karoubi envelopes

Definition A.5 Let C be a category. The *Karoubi envelope* of C (also called the *idempotent completion*) is a functor $i : C \to C^\natural$, where $C^\natural$ is pseudo-abelian, satisfying the following 2-universal property: for every Karoubian category $\mathcal{D}$, the restriction functor

$$i^* : \mathrm{Funct}(C^\natural, \mathcal{D}) \to \mathrm{Funct}(C, \mathcal{D})$$

is an equivalence of categories.

Theorem A.6 *A Karoubi envelope always exists: the functor i is fully faithful and the functor $C^\natural \to (C^\natural)^\natural$ (the Karoubi envelope of $C^\natural$) is an equivalence of categories.*

Proof One may find two proofs in [SGA4]: one in Exercise 7.5 of exposé IV and the other in Exercise 8.7.8 from exposé I. We give the first proof, which has the advantage of being constructive (the second, which uses ind-objects, is more conceptual).

Define a category $C^\natural$ whose objects are the pairs (C, p) where $C \in C$ and p is a projector of C. For another object (D, q), we define:

$$C^\natural((C, p), (D, q)) = \{f \in C(C, D) \mid f = qfp\}.$$

The composition of morphisms is induced by that of C; the identity of (C, p) is p. The functor i is defined by

$$i(C) = (C, 1_C)$$

and i is the identity on morphisms. The universal property can be easily verified, which then implies the last statement. $\square$

A.1.3 The case of additive categories

Proposition A.7 *If C is additive, $C^\natural$ is additive. Furthermore, for every object $C \in C$ and every projector p of C, the canonical morphism of $C^\natural$*

$$\mathrm{Im}(p) \oplus \mathrm{Im}(1 - p) \to C$$

is an isomorphism.

(Note that if p is a projector, then $1 - p$ is also a projector and $\mathrm{Im}(1 - p) = \mathrm{Ker}(0, p) =: \mathrm{Ker}(p)$.)

Proof By Definition [Mcl, ch. VIII, § 2], $\mathcal{C}$ is additive if its Homs are abelian groups, composition of morphisms is bilinear, and finite products are representable. These properties are clearly preserved for the construction given in the proof of Theorem A.6.

For the last statement, we check that the morphisms given by Lemma A.3 (for p and $1 - p$) make C a biproduct of $\text{Im}(p)$ and $\text{Im}(1 - p)$ in the sense of [Mcl, ch. VIII, def.].   □

A.1.4 *F*-linear structures

Definition A.8 Let F be a unital ring. An *F-linear structure* on an additive category $\mathcal{A}$ is the data of a ring homomorphism $F \to \text{End}(Id_{\mathcal{A}})$.

To be more precise, the endomorphisms of the identity functor of $\mathcal{A}$ have a natural ring structure. A homomorphism as above is the data, for all $\lambda \in F$ and all $A \in \mathcal{A}$, of an endomorphism $\lambda_A : A \to A$ which is natural in A, additive and multiplicative in λ. This amounts to giving the data of two-sided F-modules on the $\mathcal{A}(A, B)$, compatible with composition.

Lemma A.9 *Let $\mathcal{A}$ be an additive category: every F-linear structure on $\mathcal{A}$ extends canonically to its Karoubian envelope.*

Proof Exercise.   □

A.2 Monoidal categories

A.2.1 Definitions

Definition A.10 A *monoidal category* is a triple $(\mathcal{A}, \otimes, a)$ where

- $\mathcal{A}$ is a category;
- $\otimes : \mathcal{A} \times \mathcal{A} \to \mathcal{A}$ is a functor; and
- the *associativity constraint a* is a natural isomorphism

$$a_{A,B,C} : A \otimes (B \otimes C) \xrightarrow{\sim} (A \otimes B) \otimes C$$

satisfying the following coherence condition: for every $A, B, C, D \in \mathcal{A}$, the pentagonal diagram

$$A \otimes (B \otimes (C \otimes D)) \xrightarrow{a_{A,B,C \otimes D}} (A \otimes B) \otimes (C \otimes D) \xrightarrow{a_{A \otimes B,C,D}} ((A \otimes B) \otimes C) \otimes D$$

commutes.

Remark A.11 This definition can be found in Mac Lane [Mcl, ch. VII, § 1], but also in Saavedra [Saa, ch. I, 1.1.1] (up to terminology). This also applies to the following definitions. I have adopted Saavedra's terminology of *constraints*.

Mac Lane's coherence theorem [Mcl, ch. VII, § 2, th. 1] states that the above commutativity implies all other commutativities (i.e. constraints on higher associativity) imaginable.

Definition A.12 A category $\mathcal{A}$ equipped with a bifunctor $\otimes$ as above is *unital* if it is equipped with a *unit object* $\mathbf{1}$ and for every $A \in \mathcal{A}$, the natural isomorphisms (unital constraints)

$$\mathbf{1} \otimes A \xrightarrow{\sim} A \xleftarrow{\sim} A \otimes \mathbf{1}.$$

If $\mathcal{A}$ is monoidal, we ask for the "obvious" compatibilities between the associativity constraints and the unital constraints.

Lemma A.13 *Let $\mathcal{A}$ be a unital monoidal category. Then the monoid* $\mathrm{End}_{\mathcal{A}}(\mathbf{1})$ *is commutative.*

Proof We have two composition laws on $\mathrm{End}_{\mathcal{A}}(\mathbf{1})$: composition of morphisms $\circ$ and the tensor product $\otimes$. One easily checks that each law is distributive with respect to the other; a famous lemma, well known to topologists (showing that higher homotopy groups are commutative), then implies that the two laws are equal and commutative. □

Definition A.14 A *symmetric monoidal category* is a monoidal category that admits *commutativity constraints*

$$\sigma_{A,B} : A \otimes B \xrightarrow{\sim} B \otimes A$$

such that $\sigma^2 = 1$, and compatible with the associativity constraints in the "obvious" way. One can also specify a unit; one then has a unital symmetric monoidal category (a $\otimes$-category ACU in the terminology of Saavedra [Saa]).

In [Saa, ch. I] one finds infinitely many results on the various compatibilities between constraints. In particular, if $\mathcal{A}$ is monoidal, a unit is unique up to isomorphism and the unital constraints imply the other constraints.

Lemma A.15 *Let A be an object of a symmetric monoidal category $\mathcal{A}$. For every $n \geq 0$, the commutativity constraint induces a canonical homomorphism*

$$\mathfrak{S}_n \to \mathrm{Aut}_{\mathcal{A}}(A^{\otimes n})$$

where $\mathfrak{S}_n$ is the symmetric group.

Proof We reduce to the case $n = 2$ by using the Coxeter presentation of $\mathfrak{S}_n$ for the generators $\sigma_i = (i, i+1)$ $(1 \leq i < n)$. The verification of the relations

$$\sigma_i \sigma_{i+1} \sigma_i = \sigma_{i+1} \sigma_i \sigma_{i+1}$$

comes from the interaction between the associativity and commutativity constraints.   □

Remark A.16 A weaker notion than symmetry is the *braid condition*: one does not require the commutativity constraint σ to be involutive, but only that it verify the *Yang–Baxter relations* [Mcl, ch. XI]. This notion is important in the theory of quantum groups, as well as for the study of the Grothendieck–Teichmüller group introduced by Drinfeld. The action of the symmetric group is then replaced by that of the *braid group*.

A.2.2 Monoidal functors

Definition A.17 Let $\mathcal{A}$, $\mathcal{B}$ be two monoidal categories. A *monoidal functor* from $\mathcal{A}$ to $\mathcal{B}$ is a pair (F, φ) where $F : \mathcal{A} \to \mathcal{B}$ is a functor and φ is a natural isomorphism

$$\varphi_{A,B} : F(A) \otimes F(B) \xrightarrow{\sim} F(A \otimes B)$$

compatible with the associativity constraints in the "obvious" way. If $\mathcal{A}$, $\mathcal{B}$ are unital and symmetric, then we also have the notion of a unital symmetric monoidal functor.

Remark A.18 It often happens that we have a natural transformation as above which is not an isomorphism. We then speak of a pseudomonoidal functor. One sometimes also says *weak monoidal*, or *lax-monoidal*. In addition, Mac Lane says "monoidal" where I use "pseudomonoidal" and "strong monoidal" where I say "monoidal". One must check the terminology according to the author!

Finally, we say that a functor is *strictly monoidal* if φ is the identity: this terminology seems reasonably universal.

A monoidal category is said to be *strict* if its associativity constraint is equal to the identity. A statement equivalent to Mac Lane's theorem is the following.

Theorem A.19 ([Mcl, ch. XI, § 2, th. 1]) *Every monoidal category is (monoidally) equivalent to a strict monoidal category.*

Thanks to this theorem, we will not bother to write products of three or more objects with parentheses.

A.2.3 Closed monoidal categories; dualisable objects

Definition A.20 A monoidal category $\mathcal{A}$ is *closed* if the functor $\otimes$ admits a right adjoint, known as the *internal* Hom.

More precisely, the definition requires a bifunctor $\underline{\text{Hom}} : \mathcal{A}^{\text{op}} \times \mathcal{A} \to \mathcal{A}$ and isomorphisms

$$\mathcal{A}(A \otimes B, C) \simeq \mathcal{A}(A, \underline{\text{Hom}}(B, C))$$

which are natural in A, B, C.

If $\mathcal{A}$ is unital, we apply the above identity with $A = \mathbf{1}$; we find:

$$\mathcal{A}(B, C) \simeq \mathcal{A}(\mathbf{1}, \underline{\text{Hom}}(B, C))$$

which justifies the terminology "internal Hom".

Definition A.21 (Dold–Puppe) Let $\mathcal{A}$ be a unital symmetric monoidal category. An object $A \in \mathcal{A}$ is *strongly dualisable* if there exists a triple (A^*, η, ε) with $A^* \in \mathcal{A}$ (the *dual* of A) and

$$\eta : \mathbf{1} \to A \otimes A^* \text{ (unit)}, \quad \varepsilon : A^* \otimes A \to \mathbf{1} \text{ (counit)}$$

such that the compositions

$$A \xrightarrow{\eta \otimes 1_A} A \otimes A^* \otimes A \xrightarrow{1_A \otimes \varepsilon} A$$
$$A^* \xrightarrow{1_{A^*} \otimes \eta} A^* \otimes A \otimes A^* \xrightarrow{\varepsilon \otimes 1_{A^*}} A^*$$

are equal to the identity.

We say that $\mathcal{A}$ is *rigid* if every object of $\mathcal{A}$ is strongly dualisable.

Proposition A.22

(1) *If A is strongly dualisable, then the triple (A^*, η, ε) is unique up to unique isomorphism. Moreover, A^* is strongly dualisable, with dual A and with structure constants $\eta^* = \sigma \circ \eta$, $\varepsilon^* = \varepsilon \circ \sigma$ where σ is the commutativity constraint.*

(2) *Let A be strongly dualisable. Then for all $B, C \in \mathcal{A}$, we have natural isomorphisms*

$$\mathcal{A}(B \otimes A, C) \simeq \mathcal{A}(B, C \otimes A^*), \quad \mathcal{A}(B, C \otimes A) \simeq \mathcal{A}(B \otimes A^*, C).$$

(3) *If A, B are strongly dualisable with duals A^*, B^*, then $A \otimes B$ is strongly dualisable with dual $A^* \otimes B^*$.*

Proof (1) Uniqueness is left to the reader. The second statement is obtained by applying the permutation (13) to the central terms of the two compositions.

(2) By way of example, we construct the first isomorphism; in one direction:

$$\mathcal{A}(B \otimes A, C) \xrightarrow{\otimes 1_{A^*}} \mathcal{A}(B \otimes A \otimes A^*, C \otimes A^*) \xrightarrow{(1_B \otimes \eta)^*} \mathcal{A}(B, C \otimes A^*).$$

And in the other:

$$\mathcal{A}(B, C \otimes A^*) \xrightarrow{\otimes 1_A} \mathcal{A}(B \otimes A, C \otimes A^* \otimes A) \xrightarrow{(1_C \otimes \varepsilon)_*} \mathcal{A}(B \otimes A, C).$$

The axioms imply that composition in both directions is equal to the identity.

(3) Left to the reader. □

Example A.23 By taking $B = \mathbf{1}$ and C strongly dualisable, we get an isomorphism

$$\mathcal{A}(A, C) \simeq \mathcal{A}(\mathbf{1}, C \otimes A^*) \simeq \mathcal{A}(\mathbf{1}, A^* \otimes C) \simeq \mathcal{A}(C^*, A^*)$$

which we call *transposition*: we write $f \mapsto {}^t f$. This operation is involutive and satisfies

$$^t(g \circ f) = {}^t f \circ {}^t g, \quad {}^t(f \otimes g) = {}^t f \otimes {}^t g.$$

Lemma A.24 *Let $\mathcal{A}$ be a unital symmetric monoidal category, and let $A \in \mathcal{A}$ be a strongly dualisable object. Then for every unital symmetric monoidal functor $F : \mathcal{A} \to \mathcal{B}$ into another unital symmetric monoidal category, $F(A)$ is strongly dualisable.*

Proof Trivial. □

A.2.4 Categories with suspension

Definition A.25 a) A *category with suspension* is a pair $(\mathcal{A}, \Sigma_{\mathcal{A}})$ where $\mathcal{A}$ is a category and $\Sigma_{\mathcal{A}}$ is an endofunctor of $\mathcal{A}$. A *morphism* of categories with suspension $(\mathcal{A}, \Sigma_{\mathcal{A}}) \to (\mathcal{B}, \Sigma_{\mathcal{B}})$ is a pair (F, ρ) where $F : \mathcal{A} \to \mathcal{B}$ is a functor and $\eta : F \circ \Sigma_{\mathcal{A}} \Rightarrow \Sigma_{\mathcal{B}} \circ F$ is a natural isomorphism. A *2-morphism* between parallel morphisms $(F, \varphi), (G, \psi)$ is a natural transformation $u : F \Rightarrow G$ such that the diagram

$$
\begin{array}{ccc}
F\Sigma_{\mathcal{A}}(A) & \xrightarrow{\varphi_A} & \Sigma_{\mathcal{B}}F(A) \\
{\scriptstyle u_{\Sigma_{\mathcal{A}}(A)}} \downarrow & & \downarrow {\scriptstyle \Sigma_{\mathcal{B}}(u_A)} \\
G\Sigma_{\mathcal{A}}(A) & \xrightarrow{\psi_A} & \Sigma_{\mathcal{B}}G(A)
\end{array}
$$

commutes for all $A \in \mathcal{A}$. Given two categories with suspension, $(\mathcal{A}, \Sigma_{\mathcal{A}})$ and $(\mathcal{B}, \Sigma_{\mathcal{B}})$, we write

$$\underline{\mathrm{Hom}}((\mathcal{A}, \Sigma_{\mathcal{A}}), (\mathcal{B}, \Sigma_{\mathcal{B}}))$$

for the category whose objects are the morphisms from $(\mathcal{A}, \Sigma_{\mathcal{A}})$ to $(\mathcal{B}, \Sigma_{\mathcal{B}})$ and whose morphisms are the 2-morphisms.

b) If $(\mathcal{A}, \Sigma_{\mathcal{A}})$ is a category with suspension, we say that $\Sigma_{\mathcal{A}}$ is *invertible* if it is an autoequivalence.

Lemma A.26 *Let $(\mathcal{A}, \Sigma_{\mathcal{A}})$ be a category with suspension. Then there exists a morphism $\rho : (\mathcal{A}, \Sigma_{\mathcal{A}}) \to (\mathcal{A}[\Sigma_{\mathcal{A}}^{-1}], \tilde{\Sigma}_{\mathcal{A}})$ where $\tilde{\Sigma}_{\mathcal{A}}$ is invertible and 2-universal in the following sense: for every category with suspension $(\mathcal{B}, \Sigma_{\mathcal{B}})$ where $\Sigma_{\mathcal{B}}$ is invertible, the functor*

$$\rho^* : \underline{\mathrm{Hom}}((\mathcal{A}[\Sigma_{\mathcal{A}}^{-1}], \tilde{\Sigma}_{\mathcal{A}}), (\mathcal{B}, \Sigma_{\mathcal{B}})) \to \underline{\mathrm{Hom}}((\mathcal{A}, \Sigma_{\mathcal{A}}), (\mathcal{B}, \Sigma_{\mathcal{B}}))$$

is an equivalence of categories.
We may also choose $\tilde{\Sigma}_{\mathcal{A}}$ to be an automorphism of categories.

Proof Let $\mathcal{A}[\Sigma_{\mathcal{A}}^{-1}]$ be the category whose objects are pairs (A, m) for $A \in \mathcal{A}$, $m \in \mathbf{Z}$, the morphisms being given by the formula

$$\mathcal{A}[\Sigma_{\mathcal{A}}^{-1}]((A, m), (B, n)) = \varinjlim_{k \gg 0} \mathcal{A}(\Sigma_{\mathcal{A}}^{m+k}(A), \Sigma_{\mathcal{A}}^{n+k}(B))$$

where the transition morphisms are induced by $\Sigma_{\mathcal{A}}$. Composition of morphisms is defined in the obvious way. Let $\tilde{\Sigma}_A$ be the endofunctor

$$(A, m) \mapsto (A, m)[1] = (A, m+1), \quad f[1] = f$$

for the canonical identification

$$\mathcal{A}[\Sigma_{\mathcal{A}}^{-1}]((A, m), (B, n)) = \mathcal{A}[\Sigma_{\mathcal{A}}^{-1}]((A, m + 1), (B, n + 1)).$$

This is obviously an automorphism of $\mathcal{A}[\Sigma_{\mathcal{A}}^{-1}]$. We have a canonical functor

$$\rho : \mathcal{A} \to \mathcal{A}[\Sigma_{\mathcal{A}}^{-1}],$$
$$A \mapsto (A, 0),$$
$$f \mapsto f.$$

The identity $\Sigma_{\mathcal{A}}^{m+k}(\Sigma_{\mathcal{A}}(A)) = \Sigma^{m+k+1}(A)$ gives a natural isomorphism $(A, 1) \overset{\sim}{\to} (\Sigma_{\mathcal{A}} A, 0)$, that is,

$$\eta : [1] \circ \rho \overset{\sim}{\to} \rho \circ \Sigma_{\mathcal{A}}.$$

Given two morphisms $(F, \varphi), (G, \psi) : (\mathcal{A}[\Sigma_{\mathcal{A}}^{-1}], [1]) \rightrightarrows (\mathcal{B}, \Sigma_{\mathcal{B}})$, a 2-morphism $u : F \circ \rho \Rightarrow G \circ \rho$ extends uniquely to a 2-morphism $u : F \Rightarrow G$ by the formula

$$u_{(A,m)} = u_A[m].$$

This shows that ρ^* is fully faithful. Secondly, if $(F, \varphi) : (\mathcal{A}, \Sigma_{\mathcal{A}}) \to (\mathcal{B}, \Sigma_{\mathcal{B}})$ is a morphism with $\Sigma_{\mathcal{B}}$ invertible, then it extends, up to 2-isomorphism, to a morphism $(\tilde{F}, \tilde{\varphi}) : (\mathcal{A}[\Sigma_{\mathcal{A}}^{-1}], [1]) \to (\mathcal{B}, \Sigma_{\mathcal{B}})$ by the formula

$$F(A, m) = \begin{cases} \Sigma_{\mathcal{B}}^m F(A) & \text{if } m \geq 0, \\ {\Sigma'_{\mathcal{B}}}^{-m} F(A) & \text{if } m < 0, \end{cases}$$

where $\Sigma'_{\mathcal{B}}$ is a quasi-inverse of $\Sigma_{\mathcal{B}}$. This shows that ρ^* is essentially surjective. $\square$

A.2.5 Invertible objects; multiplicative localisation

Let $\mathcal{A}$ be a unital symmetric monoidal category.

Lemma A.27 *For an object $L \in \mathcal{A}$, the following conditions are equivalent:*

(i) *The endofunctor $t_L : A \mapsto A \otimes L$ is an equivalence of categories.*
(ii) *There exists an $L' \in \mathcal{A}$ and an isomorphism $\mathbf{1} \overset{\sim}{\to} L' \otimes L$.*

Proof (ii) is the special case "$\mathbf{1}$ is in the essential image of t_L" of (i). Conversely, if (i) holds, then a quasi-inverse of t_L is given by $t_{L'}$. $\square$

Definition A.28 An object L of $\mathcal{A}$ is *invertible* if it satisfies the equivalent conditions of Lemma A.27, and *quasi-invertible* if the functor t_L of the lemma is fully faithful.

Lemma A.29

(1) *For every object L of $\mathcal{A}$, we have a canonical homomorphism*

$$\mathrm{End}_{\mathcal{A}}(\mathbf{1}) \to \mathrm{End}_{\mathcal{A}}(L).$$

(2) *If L is quasi-invertible, then it is an isomorphism; in particular, the monoid $\mathrm{End}_{\mathcal{A}}(L)$ is commutative.*
(3) *If L is quasi-invertible, then for every $n \geq 2$ the action of $\mathfrak{S}_n$ on $L^{\otimes n}$ (Lemma A.15) factors through the signature.*

Proof (1) The homomorphism is induced by t_L, modulo the unit constraint.
(2) This is obvious.
(3) follows from (2) applied to $L^{\otimes n}$. $\qquad\qquad\Box$

Definition A.30 Let $L \in \mathcal{A}$. We write $\mathcal{A}[L^{-1}]$ for the category $\mathcal{A}[t_L^{-1}]$, where $(\mathcal{A}, t_L)$ is considered as a category with suspension (cf. Definition A.25 and Lemma A.26).

By construction, we have a canonical functor $\mathcal{A} \to \mathcal{A}[L^{-1}]$. It is not true in general that the symmetric monoidal structure of $\mathcal{A}$ extends to $\mathcal{A}[L^{-1}]$: by Lemma A.29 (3), a necessary condition is that the permutation (123) operates trivially on the image of $L^{\otimes 3}$ in $\mathcal{A}[L^{-1}]$. In fact, we have:

Proposition A.31 (Voevodsky [134]) *The above condition suffices.*

For an elegant proof, I refer to Riou [96, rem. 5.4 and th. 5.5].[1]
In particular:

Theorem A.32 *If L is quasi-invertible, the functor $\mathcal{A} \to \mathcal{A}[L^{-1}]$ is fully faithful, and the symmetric monoidal structure of $\mathcal{A}$ extends to $\mathcal{A}[L^{-1}]$ along this functor.*

Proof The first statement is clear, and the second follows from the previous proposition. $\qquad\qquad\Box$

---

[1] This problem does not appear explicitly in Saavedra, but seems implicit in the calculations of [Saa, ch. I, 2.6].

A.2.6 Monoidal categories and Karoubi envelopes

Proposition A.33 *Let $\mathcal{A}$ be a category. Every monoidal structure on $\mathcal{A}$ extends canonically to its Karoubi envelope. The same holds for a unital monoidal structure, or a unital symmetric monoidal structure. If $\mathcal{A}$ is closed, then so is $\mathcal{A}^\natural$.*

Proof By the universal property of Karoubi envelopes, the bifunctor

$$\mathcal{A} \times \mathcal{A} \xrightarrow{\otimes} \mathcal{A} \xrightarrow{i} \mathcal{A}^\natural$$

extends canonically to a functor from $(\mathcal{A} \times \mathcal{A})^\natural \simeq \mathcal{A}^\natural \times \mathcal{A}^\natural$. The associativity constraints, etc., extend similarly.   □

Proposition A.34 *Let $\mathcal{A}$ be rigid, symmetric monoidal and Karoubian. Let $A \in \mathcal{A}$, and let $p = p^2$ be a projector of A. Suppose that A is strongly dualisable; then $\operatorname{Im} p$ is strongly dualisable with dual $\operatorname{Im}{}^t p$ (see Example A.23).*

Proof Exercise.   □

A.2.7 The trace of an endomorphism

Definition A.35 Let $\mathcal{A}$ be a unital symmetric monoidal category, $A \in \mathcal{A}$ be a strongly dualisable object, and f be an endomorphism of A. We associate to it an element

$$\operatorname{Tr}(f) \in \operatorname{End}_{\mathcal{A}}(\mathbf{1})$$

defined as the composition

$$\mathbf{1} \xrightarrow{\eta} A \otimes A^* \xrightarrow{f \otimes 1} A \otimes A^* \xrightarrow{\sigma} A^* \otimes A \xrightarrow{\varepsilon} \mathbf{1}.$$

This is the *trace* of f. If we want to be precise, we write $\operatorname{Tr}_A(f)$. Write $\chi(A) = \operatorname{Tr}(1_A)$: this is the *Euler–Poincaré characteristic* of A.

Proposition A.36 *Let A, B be strongly dualisable objects of $\mathcal{A}$. The trace has the following properties:*

$$\operatorname{Tr}(fg) = \operatorname{Tr}(gf) \quad \textit{if } f \in \mathcal{A}(A, B), g \in \mathcal{B}(B, A);$$
$$\operatorname{Tr}(f \otimes g) = \operatorname{Tr}(f)\operatorname{Tr}(g) \quad \textit{if } f \in \operatorname{End}_{\mathcal{A}}(A), g \in \operatorname{End}_{\mathcal{A}}(B).$$

Moreover, if $f \in \mathcal{A}(A, B), g \in \mathcal{B}(B, A)$, then writing ι_{AB} for the canonical isomorphism $\mathcal{A}(\mathbf{1}, A^* \otimes B) \xrightarrow{\sim} \mathcal{A}(A, B)$, we have the following formula:

$$\operatorname{Tr}(g \circ f) = {}^t(\iota_{AB}^{-1}(f)) \circ \iota_{BA}^{-1}(g).$$

Proof Left as an exercise to the reader. □

Lemma A.37 (The trace formula) *Let* $(\mathcal{A}, A, f)$ *be as in Definition* A.35, *and let* $T : \mathcal{A} \to \mathcal{B}$ *be a unital symmetric monoidal functor. Then*

$$T(\mathrm{Tr}_{\mathcal{A}}(f)) = \mathrm{Tr}_{\mathcal{B}}(T(f)).$$

Proof Trivial. □

A.2.8 The additive case: the zeta function of an endomorphism

Definition A.38 Let $\mathcal{A}$ be a category, and let F be a unital commutative ring. Then a monoidal structure and an F-linear structure on $\mathcal{A}$ are *compatible* if $\otimes$ is bi-additive and, for any two morphisms f, g and any $\lambda \in F$, we have

$$\lambda(f \otimes g) = (\lambda f) \otimes g = f\lambda \otimes g = f \otimes \lambda g = f \otimes (g\lambda) = (f \otimes g)\lambda.$$

In this case, $\mathrm{End}_{\mathcal{A}}(\mathbf{1})$ is an F-algebra, and if $A \in \mathcal{A}$ is strongly dualisable, then the trace $\mathrm{Tr}_{\mathcal{A}} : \mathrm{End}_{\mathcal{A}}(A) \to \mathrm{End}_{\mathcal{A}}(\mathbf{1})$ is F-linear.

Examples A.39

(1) Let F be a field, and let $\mathcal{A} = \mathbf{Vec}_F$ (non-graded vector spaces) be equipped with the tensor product of vector spaces. Then $V \in \mathcal{A}$ is strongly dualisable if and only if it is finite dimensional; if $f \in \mathrm{End}(V)$, then $\mathrm{Tr}_{\mathcal{A}}(f)$ is the usual trace. To see this easily, one can proceed as in the note [12] from page 68; it may also be proved directly.

(2) Same as above, but set $\mathcal{A} = \mathbf{Vec}_F^*$, where the commutativity constraint is given by the Koszul rule: if $V, W \in \mathcal{A}$ are homogeneous of degree i, j, then for $(v, w) \in V \times W$

$$\sigma(v \otimes w) = (-1)^{ij} w \otimes v.$$

If $V \in \mathbf{Vec}_F^*$ is finite dimensional and $f \in \mathrm{End}(V)$, we then find that

$$\mathrm{Tr}_{\mathcal{A}}(f) = \sum_{i \in \mathbf{Z}} (-1)^i \mathrm{Tr}(f \mid V^i).$$

Definition A.40 Suppose that $F = \mathrm{End}_{\mathcal{A}}(\mathbf{1})$ is a $\mathbf{Q}$-algebra. Let f be an endomorphism of a strongly dualisable object A. We define its *zeta function* to be:

$$Z(f, t) = \exp\left(\sum_{n=1}^{\infty} \mathrm{Tr}_{\mathcal{A}}(f^n) \frac{t^n}{n}\right) \in F[[t]].$$

Theorem 3.65 and its proof take the following abstract form:

Theorem A.41 *Suppose that F is a field, and that there exists a symmetric monoidal functor*

$$H^* : \mathcal{A} \to \mathbf{Vec}_K^*,$$

where K is an extension of F. Then, for every strongly dualisable object $A \in \mathcal{A}$ and every endomorphism f of A, we have

$$Z(f, t) \in F(t).$$

If f is an isomorphism, then we have a functional equation of the form

$$Z({}^t f^{-1}, t^{-1}) = \det(f)(-t)^{\chi(A)} Z(f, t)$$

where ${}^t f \in \mathrm{End}(A^*)$ is the transpose of f and $\det(f)$ is a scalar; if $Z(f, t) = \frac{\prod_{i=1}^{m}(1-\alpha_i t)}{\prod_{j=1}^{n}(1-\beta_j t)}$ with $\alpha_i, \beta_j \in \bar{F}$, then we have

$$\det(f) = \left(\prod_{i=1}^{m} \alpha_i \right) \Big/ \left(\prod_{j=1}^{n} \beta_j \right).$$

Proof We may suppose that $\mathcal{A} = \mathbf{Vec}_K^*$ and $H^* = Id$. Then f is homogeneous of degree 0, which allows us to reduce to the case where A is homogeneous, say of degree i. We reason once again as in note [12] from page 68, and reduce to the case $\dim A = 1$, where f is multiplication by a scalar α. We then find

$$Z(f, t) = \begin{cases} \dfrac{1}{1 - \alpha t} & \text{if } i \text{ is even,} \\ 1 - \alpha t & \text{if } i \text{ is odd.} \end{cases}$$

The functional equation, with the value of $\det(f)$, can be found by observing that $\chi(A) = (-1)^i$. $\qquad\qquad\square$

In [60, th. 3.2 and rem. 3.3] one may find weaker conditions that still imply the conclusions of this theorem.

Appendix B

Triangulated categories, derived categories, and perfect complexes

B.1 Localisation

Reference: Gabriel–Zisman [GZ].

B.1.1 Categories of fractions

If A is a commutative ring, and S is a (multiplicative) subset of A, one knows how to define $S^{-1}A$, the localisation of A with respect to S [BAC2]: it is equipped with a homomorphism $A \rightarrow S^{-1}A$ which is universal for homomorphisms making the elements of S invertible.

If A is an integral domain, then the construction of $S^{-1}A$ is easy: it is the subring of the field of fractions of A generated by A and S^{-1}. In general, one must be more careful with the construction [BAC2, § 2, n° 1, def. 2].

When A is not commutative (but still unital), the universal property given above always has a solution, even if it is not easy to find a reference for this in the literature, other than the case where S satisfies the "Ore condition".

Modulo set-theoretic problems, this situation generalises[1] to an arbitrary category as follows:

Definition B.1 Let C be a category, and let S be a set of morphisms of C. A functor $F : C \rightarrow D$ *inverts* S if $F(s)$ is an isomorphism of D for every $s \in S$. A *localisation* of C with respect to S is a functor $F_u : C \rightarrow C'$ satisfying the following 2-universal property: for every category D, the functor

$$\text{Funct}(C', D) \xrightarrow{F_u^*} \text{Funct}(C, D)$$

is fully faithful, and its essential image consists of the functors inverting S.

---

[1] See example B.3.

Theorem B.2 ([GZ, ch. I, 1.1 and 1.2]) *If C is small, then it admits a localisation. More precisely, there exists a category $S^{-1}C$ having the same objects as C and a functor $P_S : C \to S^{-1}C$ which is the identity on objects, satisfying the above 2-universal property.*

(A category C is *small* if the class of its objects, $Ob(C)$, is a set, and if $C(c, d)$ is a set for every $(c, d) \in Ob(C)^2$. In Grothendieck's language of universes, for a given universe $\mathcal{U}$, we say that a set is $\mathcal{U}$-small if it belongs to $\mathcal{U}$.)

Example B.3 Let M be a monoid. We associate to it a category with one object, denoted BM. If S is a subset of M, the category $S^{-1}BM$ is of the form $B(S^{-1}M)$ where $S^{-1}M$ is the *monoid of fractions of M with respect to S*. If M is a ring, we get the ring of fractions of M.

Theorem B.4 ([GZ, ch. I, 1.3]) *Let $G : C \to D$ be a functor with right adjoint $D : D \to C$. Write $\varepsilon : GD \to Id_D$ for the counit of this adjunction. Then the following conditions are equivalent:*

(i) *ε is an isomorphism of functors;*
(ii) *D is fully faithful;*
(iii) *G is a localisation relative to $S = \{s \in Fl(C) \mid G \text{ inverts } s\}$;*
(iv) *for every category $\mathcal{E}$, the functor*

$$\text{Funct}(D, \mathcal{E}) \xrightarrow{G^*} \text{Funct}(C, \mathcal{E})$$

is fully faithful.

B.1.2 Calculus of fractions

If M is a noncommutative monoid, the monoid $S^{-1}M$ is in general very difficult to describe when $S \subset M$ is arbitrary: a generic element of $S^{-1}M$ is of the form $s_1^{-1}m_1s_2^{-1}m_2 \ldots s_r^{-1}m_r$, where $s_i \in S$, $m_i \in M$, the integer r being *a priori* unbounded and the relations between the generators being very non-explicit. The situation is better when we have a *calculus of fractions*: this is the Ore condition when M is "without torsion". This situation generalises to the case of an arbitrary category, and gives rise to:

Definition B.5 ([GZ, ch. I, 2.2]) Let C be a category, and let $S \subset Ar(C)$ be a class of morphisms. We say that S *admits a calculus of left fractions* if the following conditions are satisfied:

Multiplicativity: S contains the identity morphisms of objects of C and is stable under composition.

The Ore condition: For every diagram $X' \xleftarrow{s} X \xrightarrow{u} Y$ where $s \in S$, there exists a commutative square

$$
\begin{array}{ccc}
X & \xrightarrow{\;u\;} & Y \\
{\scriptstyle s}\downarrow & & \downarrow{\scriptstyle s'} \\
X' & \xrightarrow{\;u'\;} & Y'
\end{array}
$$

with $s' \in S$.

Simplification: If $X \underset{g}{\overset{f}{\rightrightarrows}} Y$ are parallel morphisms, and $s \in S$ is such that $fs = gs$, there exists a $t \in S$ such that $tf = tg$.

We have a dual definition for the calculus of right fractions.

If S admits a calculus of left fractions, then the Ore condition implies that every arrow of $S^{-1}C$ can be written $s^{-1}f$, where $f \in Ar(C)$ and $s \in S$. This also provides a composition rule for two such arrows; finally, the simplification condition tells us when two pairs (f, s) and (f', s') define the same arrow $s^{-1}f$: see [GZ, ch. I, 2.3] for more details.

Example B.6 Let $\mathcal{A}$ be an abelian category. A subcategory $\mathcal{B} \subset \mathcal{A}$ is called a *Serre subcategory* if it is full, and if given a short exact sequence $0 \to A' \to A \to A'' \to 0$, we have $A \in \mathcal{B} \iff A', A'' \in \mathcal{B}$. We associate to $\mathcal{B}$ the class of morphisms

$$
S_\mathcal{B} = \{s \mid \mathrm{Ker}\, s, \mathrm{Coker}\, s \in \mathcal{B}\}.
$$

Then $S_\mathcal{B}$ admits a calculus of left and right fractions, and the category $\mathcal{A}/\mathcal{B} := S_\mathcal{B}^{-1}\mathcal{A}$ is abelian: this is the *Serre quotient* of $\mathcal{A}$ by $\mathcal{B}$. Thus, we may say that the first example of localisation of categories appearing in the literature is that of Serre in [107] (where $\mathcal{A} = \{$abelian groups$\}$ and $\mathcal{B} =$, e.g., $\{$abelian groups with finite p-primary torsion$\}$). See also [108, nos 56–60].

Another example, below, will be that of a thick subcategory of a triangulated category.

Theorem B.7 ([GZ, ch. I, 4.1, prop.]) *Let C be a small category, and let $S \subset Ar(C)$ be a set of arrows admitting a calculus of left fractions. Then the localisation functor*

$$
P_S : C \to S^{-1}C
$$

admits a right adjoint if and only if the following condition holds: for every object c ∈ C, there exists an object d and a morphism s : c → d such that

(i) $P_S(s)$ *is invertible;*

(ii) *for every t : x → y in S, the map $t^* : C(y, d) → C(x, d)$ is bijective.*

If this condition is satisfied, then c ↦ d defines the desired right adjoint, and the arrow s corresponds to the unit of the adjunction.

B.2 Triangulated categories and derived categories

References: Verdier's thesis [Ver] and état 0 [SGA4$\frac{1}{2}$, C.D.], Residues and Duality [Har1, ch. 1], . . .

B.2.1 The cone of a morphism of complexes

Let $\mathcal{A}$ be an additive category: write $C(\mathcal{A})$ for the category of cochain complexes

$$\cdots \to C^{n-1} \xrightarrow{d^{n-1}} C^n \xrightarrow{d^n} C^{n+1} \to \cdots$$

of $\mathcal{A}$.

Definition B.8 Let $f : C \to D$ be a morphism in $C(\mathcal{A})$. The *cone* of f is the complex E where

$$E^n = D^n \oplus C^{n+1}$$

and the differential $d_E^n : E^n \to E^{n+1}$ is given by the matrix

$$d_E^n = \begin{pmatrix} d_D^n & f^{n+1} \\ 0 & -d_C^{n+1} \end{pmatrix}.$$

We have morphisms of complexes

$$D \xrightarrow{g} E \xrightarrow{h} C[1]$$

where $(C[1])^n = C^{n+1}, d_{C[1]}^n = -d_C^{n+1}$:

$$g(d) = (d, 0); \quad h(d, c) = c.$$

In [Ver, ch. I, prop. 3.1.3] one finds a description of this construction as arising from an adjunction.

B.2.2 Homotopies

Definition B.9 Let $C \overset{f}{\underset{g}{\rightrightarrows}} D$ be parallel morphisms of $C(\mathcal{A})$. A *homotopy from f to g* is a graded homomorphism of degree -1:

$$H^n : C^n \to D^{n-1}$$

such that $g - f = d_D H + H d_C$.

If there exists a homotopy between f and g, we say that they are *homotopic*.

Lemma B.10 *Homotopy is an equivalence relation and is compatible with addition and composition of morphisms.* □

Lemma B.11 *Let $f : C \to D$ be a morphism in $C(\mathcal{A})$ with cone E, and let g, h be as in Definition B.8. Then $h \circ g = 0$ and $g \circ f$, $f[1] \circ h$ are homotopic to 0.*

Proof A homotopy H_1 from 0 to $g \circ f$ is given by

$$H_1(c) = (0, c)$$

and a homotopy H_2 from 0 to $f[1] \circ h$ is given by

$$H_2(d, c) = d.$$ □

Definition B.12 Let $K(\mathcal{A})$ be the category whose objects are those of $C(\mathcal{A})$, and whose morphisms are given by

$$K(\mathcal{A})(C, D) = C(\mathcal{A})(C, D)/ \sim_h,$$

where $\sim_h$ is the homotopy equivalence relation (cf. Lemma B.10). This is the *homotopy category* of $\mathcal{A}$. If $f : C \to D$ is a morphism of $C(\mathcal{A})$, we associate to it the triangle from Definition B.8

$$t(f) : C \overset{f}{\to} D \overset{g}{\to} E \overset{h}{\to} C[1],$$

mapped to $K(\mathcal{A})$.

B.2.3 Triangulated categories

Definition B.13 a) A **Z**-*category* is a category $\mathcal{T}$ equipped with an autoequivalence of categories $X \mapsto X[1]$; if it is an automorphism of categories, then we say it is a *strict* **Z**-*category*. One calls this autoequivalence the *shift, suspension* or *translation* automorphism.

b) A *triangle* of $\mathcal{T}$ is a diagram

$$X \to Y \to Z \to X[1]$$

often abbreviated as

$$X \to Y \to Z \xrightarrow{+1}.$$

A *morphism of triangles* is the obvious notion.

Another notation that is common in the literature, explaining the terminology, is

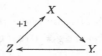

Definition B.14 (Verdier [Ver, ch. II, def. 1.1.1]) A *triangulated category* is the data of a strict additive **Z**-category $\mathcal{T}$ and a class of triangles, called *exact triangles*, with the following properties:

TR1 Every triangle that is isomorphic to an exact triangle is exact. For every object X of $\mathcal{T}$, the triangle $X \xrightarrow{1_X} X \to 0 \to X[1]$ is exact. Every morphism $u : X \to Y$ of $\mathcal{T}$ is contained in an exact triangle $X \xrightarrow{u} Y \xrightarrow{v} Z \xrightarrow{w} X[1]$.

TR2 A triangle $X \xrightarrow{u} Y \xrightarrow{v} Z \xrightarrow{w} X[1]$ is exact if and only if the triangle $Y \xrightarrow{v} Z \xrightarrow{w} X[1] \xrightarrow{-u[1]} Y[1]$ is exact.

TR3 Every commutative diagram

$$
\begin{array}{ccccccc}
X & \xrightarrow{u} & Y & \xrightarrow{v} & Z & \xrightarrow{w} & X[1] \\
\downarrow{\scriptstyle f} & & \downarrow{\scriptstyle g} & & & & \\
X' & \xrightarrow{u'} & Y' & \xrightarrow{v'} & Z' & \xrightarrow{w'} & X'[1]
\end{array}
$$

with exact rows can be completed to a commutative diagram

$$
\begin{array}{ccccccc}
X & \xrightarrow{u} & Y & \xrightarrow{v} & Z & \xrightarrow{w} & X[1] \\
\downarrow{\scriptstyle f} & & \downarrow{\scriptstyle g} & & \downarrow{\scriptstyle h} & & \downarrow{\scriptstyle f[1]} \\
X' & \xrightarrow{u'} & Y' & \xrightarrow{v'} & Z' & \xrightarrow{w'} & X'[1].
\end{array}
$$

TR4 (the "octahedral axiom") For every commutative triangle

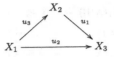

and every triple of exact triangles

$$X_1 \xrightarrow{u_2} X_2 \xrightarrow{v_3} Z_3 \xrightarrow{w_3} X_1[1]$$
$$X_2 \xrightarrow{u_1} X_3 \xrightarrow{v_1} Z_1 \xrightarrow{w_1} X_2[1]$$
$$X_1 \xrightarrow{u_3} X_3 \xrightarrow{v_2} Z_2 \xrightarrow{w_2} X_1[1]$$

there exist morphisms $m_1 : Z_3 \to Z_2$ and $m_3 : Z_2 \to Z_1$ such that $(1_{X_1}, u_1, m_1)$ and $(u_3, 1_{X_3}, m_3)$ are morphisms of triangles, and the triangle

$$Z_3 \xrightarrow{m_1} Z_2 \xrightarrow{m_1} Z_1 \xrightarrow{v_3[1]w_1} Z_3[1]$$

is exact.

See [13, 1.1.7–1.1.14] for comments on the octahedral axiom.

Theorem B.15 ([Ver, ch. II, prop. 1.3.2]) *Let $\mathcal{A}$ be an additive category. The category $K(\mathcal{A})$ from Definition B.12, equipped with the triangles $t(f)$, is triangulated.*

Definition B.16 a) Let $\mathcal{T}, \mathcal{T}'$ be triangulated categories. A *triangulated functor* from $\mathcal{T}$ to $\mathcal{T}'$ is an additive functor $F : \mathcal{T} \to \mathcal{T}'$ equipped with natural commutation isomorphisms between the shift functors and which takes exact triangles to exact triangles.
b) Let $\mathcal{T}$ be a triangulated category, and $\mathcal{A}$ be an abelian category. A *homological functor from $\mathcal{T}$ to $\mathcal{A}$* is a functor $H : \mathcal{T} \to \mathcal{A}$ such that, for every exact triangle $X \to Y \to Z \xrightarrow{+1}$ of $\mathcal{T}$, the sequence (in $\mathcal{A}$)

$$HX \to HY \to HZ \to H(X[1])$$

is exact (H is automatically additive: [Ver, ch. II, rem. 1.2.7]).
A *cohomological functor from $\mathcal{T}$ to $\mathcal{A}$* is a homological functor from $\mathcal{T}^{\mathrm{op}}$ to $\mathcal{A}$.

Definition B.17 Let $\mathcal{T}, \mathcal{T}'$ be triangulated categories, and let $F_1, F_2 : \mathcal{T} \rightrightarrows \mathcal{T}'$ be parallel triangulated functors. A natural transformation $u : F_1 \Rightarrow F_2$ is *triangulated* if it commutes with shifts in the following way: for every $X \in \mathcal{T}$, the diagram

$$
\begin{array}{ccc}
F_1(X[1]) & \xrightarrow{\;\sim\;} & F_1(X)[1] \\
{\scriptstyle u_{X[1]}} \downarrow & & {\scriptstyle u_X[1]} \downarrow \\
F_2(X[1]) & \xrightarrow{\;\sim\;} & F_2(X)[1]
\end{array}
$$

commutes, where the horizontal isomorphisms come from the commutation with shifts included in the structure of F_1 and F_2.

Theorem B.18 ([Ver, ch. II, prop. 1.2.1]) *Let $\mathcal{T}$ be a triangulated category, and Ab be the category of abelian groups. For every object $X \in \mathcal{T}$, the functor $\mathcal{T}(X, -) : \mathcal{T} \to$ Ab is homological and the functor $\mathcal{T}(-, X) : \mathcal{T}^{\mathrm{op}} \to$ Ab is cohomological.*

Remarks B.19

(1) In [Ver], Verdier uses the terminology of "distinguished triangles", "exact functors", and "cohomological functors" respectively for what we call "exact triangles", "triangulated functors", and "homological functors". The terminology chosen here has been gradually adopted: "exact triangle" reflects better the fact that such a triangle generates a long exact sequence, and "exact functor" is particularly unfortunate since the usual triangulated functors are total derived functors from non-exact functors between abelian categories! Similarly, it seems preferable to reserve "homological" for covariant and "cohomological" for contravariant functors.

(2) The hypothesis that the shift functor be an automorphism of $\mathcal{T}$ excludes the topologists's stable homotopy category from our examples of triangulated categories: the suspension functor is only an autoequivalence of categories. This is however an inessential difference. I have chosen to keep Verdier's hypothesis in order to avoid possible unexpected issues.

B.2.4 The Verdier quotient

In this section, I (hypocritically) skip set-theoretic issues.

Definition B.20 a) Let $\mathcal{T}$ be a triangulated category. A subcategory $\mathcal{S} \subset \mathcal{T}$ is *triangulated* if it is full, stable under shifts, and if for every morphism $f :$ $X \to Y$ of $\mathcal{S}$, there exists an exact triangle $X \xrightarrow{f} Y \xrightarrow{g} Z \xrightarrow{+1}$ of $\mathcal{T}$ with $Z \in \mathcal{S}$. Thus the inclusion functor is triangulated.

b) $\mathcal{S}$ is *thick*[2] if it is moreover stable under direct summands.

Let $\mathcal{S} \subset \mathcal{T}$ be a triangulated subcategory. We associate to it the class of morphisms

$$S(\mathcal{S}) = \{s : X \to Y \mid X, Y \in \mathcal{T}; \text{ there is an exact triangle}$$

$$X \xrightarrow{s} Y \to Z \xrightarrow{+1} \text{ with } Z \in \mathcal{S}\}.$$

Theorem B.21 ([Ver, ch. II, prop. 2.1.8 and th. 2.2.6]) *The class $S(\mathcal{S})$ admits a calculus of left and right fractions. Write $\mathcal{T}/\mathcal{S}$ for the localised category: it possesses a unique triangulated structure such that the localisation functor $P : \mathcal{T} \to \mathcal{T}/\mathcal{S}$ is triangulated: this is the structure in which the exact triangles are images of exact triangles of $\mathcal{T}$. The category $\mathcal{T}/\mathcal{S}$ is the* Verdier quotient *of $\mathcal{S}$ by $\mathcal{T}$.*

Remark B.22 The category $\mathcal{S}^{\natural} = \operatorname{Ker} P$ is the smallest thick subcategory of $\mathcal{T}$ containing $\mathcal{S}$: its objects are the direct summands of the objects of $\mathcal{S}$. For this reason, one sometimes limits to localising with respect to thick subcategories (which Verdier did not do).

Definition B.23 Let $\mathcal{C}$ be a class of objects in $\mathcal{T}$. The *right orthogonal* (resp. *left orthogonal*) of $\mathcal{C}$ is

$$\mathcal{C}^{\perp} = \{X \in \mathcal{T} \mid \mathcal{T}(\mathcal{C}, X) = 0\}$$

(resp.

$$^{\perp}\mathcal{C} = \{X \in \mathcal{T} \mid \mathcal{T}(X, \mathcal{C}) = 0\}).$$

If $\mathcal{C}$ is stable under translation, these are thick subcategories of $\mathcal{T}$.

The following theorem translates Theorem B.7 into the triangulated context:

Theorem B.24 ([Ver, ch. II, prop. 2.3.3]) *Let $\mathcal{S}$ be a triangulated subcategory of $\mathcal{T}$. The localisation functor $P : \mathcal{T} \to \mathcal{T}/\mathcal{S}$ admits a right adjoint (resp. left adjoint) if and only if its restriction to $\mathcal{S}^{\perp}$ (resp. to $^{\perp}\mathcal{S}$) is essentially surjective.*

---

[2] Verdier says *thick* in [SGA4$\frac{1}{2}$, C.D., § 2, def. 1.1] and *saturated* in [Ver, ch. II, 2.1.6]: these two notions are equivalent.

B.2.5 Derived categories

Definition B.25 a) Let $\mathcal{A}$ be an abelian category, and let $C, D \in C(\mathcal{A})$. A morphism $f : C \to D$ is a *quasi-isomorphism* if $H^*(f) : H^*(C) \to H^*(D)$ is an isomorphism. An object C is *acyclic* if $H^*(C) = 0$, i.e. if it is quasi-isomorphic to 0.

b) We write $D(\mathcal{A})$ for the Verdier quotient $K(\mathcal{A})/\mathcal{S}$, where $\mathcal{S}$ is the thick subcategory of acyclic complexes. This is the *derived category* of $\mathcal{A}$.

Remark B.26 We have "bounded" versions of the categories $K(\mathcal{A})$ and $D(\mathcal{A})$:

$$K^+(\mathcal{A}) = \{C \in K(\mathcal{A}) \mid C^n = 0 \text{ for } n \ll 0\},$$
$$K^-(\mathcal{A}) = \{C \in K(\mathcal{A}) \mid C^n = 0 \text{ for } n \gg 0\},$$
$$K^b(\mathcal{A}) = K^+(\mathcal{A}) \cap K^-(\mathcal{A})$$

and similarly for $D^\#(\mathcal{A})$ ($\# \in \{+, -, b\}$).

Proposition B.27 *Let $K_P(\mathcal{A})$ (resp. $K_I(\mathcal{A})$) be the full subcategory of $K(\mathcal{A})$ formed of complexes whose terms are projective (resp. injective): these are thick subcategories of $K(\mathcal{A})$.*

(1) *The compositions*

$$K_P(\mathcal{A}) \to K(\mathcal{A}) \to D(\mathcal{A}), \quad K_I(\mathcal{A}) \to K(\mathcal{A}) \to D(\mathcal{A})$$

are fully faithful.

(2) *If every object of $\mathcal{A}$ is a quotient of a projective object (resp. injects into an injective object), then the essential image of the first (resp. second) composition contains $D^-(\mathcal{A})$ (resp. $D^+(\mathcal{A})$).*

Remark B.28 The above proposition is the derived version of injective and projective resolutions, which allow us to do calculations in homological algebra. In the 60s and 70s, people paid a lot of attention to questions of "boundedness" and to the pathologies that threatened to appear if these conditions were removed. This situation was revolutionised in the 1990s by Spaltenstein [120], and then Bökstedt–Neeman [14] who introduced methods allowing one to work in the same way with unbounded derived categories. A fundamental step was the paper of Thomason–Trobaugh, which identified perfect complexes with compact objects of the derived category of quasicoherent sheaves over a reasonable scheme [133]. The importance of unbounded derived categories then grew more with further work of Neeman [Nee]: for example, if $\mathcal{A}$ is an abelian category admitting infinite exact direct sums, then $D(\mathcal{A})$ is stable under infinite direct sums, which is not the case for $D^+(\mathcal{A})$ or $D^-(\mathcal{A})$, and even less so for $D^b(\mathcal{A})$.

B.2.6 K_0 of abelian and triangulated categories

If $\mathcal{A}$ is a (small) abelian category, its *Grothendieck group* $K_0(\mathcal{A})$ is the abelian group defined by generators and relations:

Generators: $[A]$, where A describes the isomorphism classes of objects of $\mathcal{A}$.
Relations: $[A] = [A']+[A'']$ for every short exact sequence $0 \to A' \to A \to A'' \to 0$.

If $\mathcal{T}$ is a (small) triangulated category, we define $K_0(\mathcal{T})$ in the same way, with relations coming from exact triangles.

If $D^b(\mathcal{A})$ is the bounded derived category of an abelian category $\mathcal{A}$, then the embedding $A \mapsto A[0]$ of $\mathcal{A}$ in $D^b(\mathcal{A})$ induces an *isomorphism*

$$K_0(\mathcal{A}) \xrightarrow{\sim} K_0(D^b(\mathcal{A}))$$

[SGA5, exp. VIII, § 4].

B.2.7 Derived functors

Let $F : \mathcal{A} \to \mathcal{B}$ be a left exact functor between abelian categories. If $\mathcal{A}$ has enough injectives in the sense of Proposition B.27 (2), then we define the *right derived functors* of F by the formula

$$R^q F(A) = H^q(I^*),$$

where I^* is an injective resolution of A. The idea underlying the concept of derived categories is to preserve the information that is contained in I^* but lost in the passage to cohomology. This leads to the notion of total derived functors.

Definition B.29 A functor $F : \mathcal{A} \to \mathcal{B}$ as above is said to be *derivable on the right* if there exists a triangulated functor $RF : D(\mathcal{A}) \to D(\mathcal{B})$ and a natural transformation

$$\varphi_A : F(A)[0] \to RF(A)$$

(where $B[0] \in D(\mathcal{B})$ denotes the object B of $\mathcal{B}$ considered as a complex concentrated in degree 0) having the following universal property: for every triangulated functor $G : D(\mathcal{A}) \to D(\mathcal{B})$ equipped with a natural transformation $\psi : F[0] \Rightarrow G$, there exists a unique natural transformation $\theta : RF \Rightarrow G$ such that $\psi = \theta \circ \varphi$.

Variant: we replace D by D^+.

This notion is a special case of Kan extensions [Mcl, X.3].

Proposition B.30 *If $\mathcal{A}$ has enough injectives, then RF is defined on $D^+(\mathcal{A})$; it extends to $D(\mathcal{A})$ [at least] if $R^q F = 0$ for $q \gg 0$. (We then say that F has finite cohomological dimension.)*

Theorem B.31 *Let $\mathcal{A} \xrightarrow{F} \mathcal{B} \xrightarrow{G} \mathcal{C}$ be left exact functors between abelian categories. Suppose that $\mathcal{A}$ and $\mathcal{B}$ have enough injectives, and that F transforms every injective object of $\mathcal{A}$ into a G-acyclic object of $\mathcal{B}$ (i.e. an object B such that $R^q G(B) = 0$ for all $q > 0$). Then there is a canonical isomorphism*

$$R(G \circ F) \simeq RG \circ RF$$

on $D^+(\mathcal{A})$ (and on $D(\mathcal{A})$, under the hypothesis of finite cohomological dimension).

This theorem replaces[3] and generalises the spectral sequence for composed functors

$$R^p G R^q F(A) \Rightarrow R^{p+q}(G \circ F)(A).$$

B.2.8 Derived and adjoint functors

What we have just discussed for right derivation applies just as well to left derivation, *mutatis mutandis*. What happens in the case of adjoint functors?

Here is an example of an answer.

Proposition B.32 *Let $\mathcal{A}, \mathcal{B}$ be two abelian categories, and let $G : \mathcal{A} \to \mathcal{B}$, $D : \mathcal{B} \to \mathcal{A}$ be a pair of adjoint functors (G is left adjoint to D): so G is right exact and D is left exact. Suppose that $\mathcal{B}$ has enough injectives and $\mathcal{A}$ has enough projectives. Then the total derived functors $LG : D^-(\mathcal{A}) \to D^-(\mathcal{B})$, $RD : D^+(\mathcal{B}) \to D^+(\mathcal{A})$ exist and their restrictions to $D^b(\mathcal{A})$ and $D^b(\mathcal{B})$ are mutually adjoint.*

For a generalisation to a very abstract setting, consult [65].

---

[3] At least for formal reasonings! It is difficult to avoid spectral sequences if one wishes to do concrete calculations...

B.2.9 Examples: $\overset{L}{\otimes}$ and RHom

Definition B.33 Let $\mathcal{A}$ be an abelian category equipped with a symmetric monoidal product $\otimes$, which we suppose to be *right exact*.

a) An object $A \in \mathcal{A}$ is *flat* if the functor $A \otimes -$ is exact.
b) An *internal Hom* is a right adjoint Hom to $\otimes$ (if it exists). An object $A \in \mathcal{A}$ is *coflat* if Hom$(-, A)$ is exact.

Proposition B.34 *Let* $(\mathcal{A}, \otimes)$ *be as above, and suppose that every object of* $\mathcal{A}$ *is the quotient of a flat object (we say that* $\mathcal{A}$ *has enough flat objects). Then the functor*

$$\otimes : \mathcal{A} \times \mathcal{A} \to \mathcal{A}$$

derives to give

$$\overset{L}{\otimes} : D^-(\mathcal{A}) \times D^-(\mathcal{A}) \to D^-(\mathcal{A}).$$

Suppose that $\mathcal{A}$ *is closed and admits enough coflat objects. Then* Hom *can be derived as*

$$\text{RHom} : D^-(\mathcal{A}) \times D^+(\mathcal{A}) \to D^+(\mathcal{A}).$$

Furthermore, the restrictions of $\overset{L}{\otimes}$ *and* RHom *to* $D^b(\mathcal{A}) \times D^b(\mathcal{A})$ *are mutually adjoint.*

In practice, we take $\mathcal{A}$ to be the category of sheaves of Λ-modules on a site X, where Λ is a sheaf of rings on X: then $\mathcal{A}$ has enough flat objects, and furthermore every injective object is coflat. The above proposition therefore applies.

B.3 Perfect complexes

Definition B.35 Let Λ be a unital ring and let $\mathcal{A}$ be the abelian category of left Λ-modules. A complex $C \in C(\mathcal{A})$ is *perfect* if it is isomorphic in $D(\mathcal{A})$ to a bounded complex of projective Λ-modules of finite type.

Definition B.36 Let (S, Λ) be a locally ringed site, and let $\mathcal{A}$ be the abelian category of left Λ-modules.

a) A complex $C \in C(\mathcal{A})$ is *strictly perfect* if it is a bounded complex of locally free Λ-modules of finite type.

b) An object $C \in D(\mathcal{A})$ is *perfect* if there exists a covering (U_i) of S such that $C_{|U_i}$ is isomorphic to a strictly perfect complex for all i.

The following theorem was first proved by Thomason–Trobaugh in a particular case [133, th. 2.4.3]; the general case was found by Amnon Neeman [87].

Theorem B.37 *Suppose that* $(S, \Lambda) = (X, \mathcal{O}_X)$ *for a separated quasicompact scheme* X. *Then an object* C *of* $D(\mathcal{A})$ *is a perfect complex if and only if it is compact, that is to say, if the functor* $D \mapsto \mathrm{Hom}(C, D)$ *commutes with infinite direct sums.*

I do not know whether this statement remains true in the case of the perfect complexes of [SGA5, Exp. XV] that are used in § 5.3.8; in any case, there are examples of quasiseparated quasicompact algebraic stacks on which certain perfect complexes are not compact, cf. [46, § 4] (I thank Neeman for this reference).

Appendix C
List of exercises

Chapter 1

Chapter 2

Chapter 3

Chapter 4

Chapter 5

Chapter 6

Bibliography

Books

[Ami] Y. Amice *Les nombres p-adiques*, PUF, 1975.

[And] Y. André *Une introduction aux motifs: motifs purs, motifs mixtes, périodes*, Panoramas et synthèses **17**, SMF, 2004.

[AT] E. Artin, J. Tate *Class field theory*, Benjamin, 1968; revised edition, AMS, 2009.

[Ayo] J. Ayoub *Les six opérations de Grothendieck et le formalisme des cycles évanescents dans le monde motivique*, Astérisque **314–315**, SMF, 2007.

[BS] Z.I. Borevič, I.R. Šafarevič *Théorie des nombres*, translated from Russian by Myriam and Jean-Luc Verley, Gauthier-Villars, 1967.

[BLR] S. Bosch, W. Lütkebohmert and M. Raynaud *Néron models*, Erg. Math. u. Grenzgebiete (3) **21**, Springer, 1990.

[BA3] N. Bourbaki *Algèbre*, ch. 3: algèbres tensorielles, algèbres extérieures, algèbres symétriques, Hermann, 1970.

[BAC2] N. Bourbaki *Algèbre commutative*, ch. 2: localisation, Masson, 1985.

[BLie2] N. Bourbaki *Groupes et algèbres de Lie*, ch. 2: algèbres de Lie libres, Masson, 1986.

[CF] J.W.S. Cassels, A. Fröhlich (eds.) *Algebraic number theory*, Proceedings of an instructional conference organized by the London Mathematical Society (a NATO advanced study institute) with the support of the International Mathematical Union, Acad. Press, 1967.

[Con] B. Conrad *Grothendieck duality and base change*, Lect. Notes in Math. **1750**, Springer, 2000.

[Dav] H. Davenport *Multiplicative number theory* (third edition), Grad. Texts in Math. **74**, Springer, 2000.

[DMOS] P. Deligne, J. Milne, A. Ogus, W. Shih *Hodge cycles, motives, and Shimura varieties*, Lect. Notes in Math. **900**, Springer, 1982.

[Eic] M. Eichler *Introduction to the theory of algebraic numbers and functions*, translated from German by George Striker. Pure and Applied Mathematics **23**, Acad. Press, 1966, xiv+324 pp.

[Ell] W. J. Ellison (in collaboration with M. Mendès-France) *Les nombres premiers*, Hermann, 1975.

197

[Ful]   W. Fulton *Intersection theory*, Springer (new ed.), 1998.

[GZ]   P. Gabriel, M. Zisman *Calculus of fractions and homotopy theory*, Springer, 1967.

[Har1] R. Hartshorne *Residues and duality*, Lecture notes of a seminar on the work of A. Grothendieck, given at Harvard 1963/64. With an appendix by P. Deligne. Lect. Notes in Math. **20**, Springer, 1966.

[Har2] R. Hartshorne *Algebraic geometry*, Springer, 1977.

[Lan1] S. Lang *Abelian varieties*, Interscience publ., 1959, reprint: Springer, 1983.

[Lan2] S. Lang *Algebraic number theory*, Springer, 1970.

[LS]   B. Le Stum *Rigid cohomology*, Cambridge University Press, 2007.

[Mcl]   S. Mac Lane *Categories for the working mathematician*, 2nd ed., Springer, 1998.

[Mil]   J. Milne *Étale cohomology*, Princeton University Press, 1980.

[Mum]   D. Mumford *Abelian varieties*, TIFR, 1970. Reissue 2008: Hindustan Book Agency.

[Nee]   A. Neeman *Triangulated categories*, Ann. Math. Studies **148**, Princeton University Press, 2001.

[Saa]   N. Saavedra Rivano *Catégories tannakiennes*, Lect. Notes in Math. **265**, Springer, 1972.

[Ser1] J.-P. Serre *Représentations linéaires des groupes finis*, Herrmann, 1967.

[Ser2] J.-P. Serre *Corps locaux*, 2nd ed., Hermann, 1968.

[Ser3] J.-P. Serre *Cours d'arithmétique*, Collection Sup, P.U.F., Paris, 1970.

[Ser4] J.-P. Serre *Algèbre locale; multiplicités*, 3rd ed., Lect. Notes in Math. **11**, Springer, 1975–1997.

[ST]   G. Shimura, Y. Taniyama *Complex multiplication of abelian varieties and its applications to number theory*, Publ. Math. Soc. Japan **6**, 1961.

[Ver]   J.-L. Verdier *Des catégories dérivées des catégories abéliennes*, Astérisque **239** (1996).

[Wei1] A. Weil *Foundations of algebraic geometry*, AMS Coll. **XXIX** (1946), 2nd ed. 1962.

[Wei2] A. Weil *Théorie élémentaire des correspondances sur une courbe*, in *Sur les courbes algébriques et les variétés qui s'en déduisent*, Hermann, 1948.

[Wei3] A. Weil *Variétés abéliennes et courbes algébriques*, Hermann, 1948.

Acronyms

[EGA4$_3$] A. Grothendieck *Éléments de géométrie algébrique* (with A. Dieudonné), Ch. IV: *Étude locale des schémas et des morphismes de schémas* (*troisième partie*), Publ. Math. IHÉS **28** (1966), 5–255.

[SGA1] Séminaire de géométrie algébrique du Bois-Marie 1960–61 (SGA 1): Revêtements étales et groupe fondamental, new ed., Documents mathématiques **3**, SMF, 2003.

[SGA4] Séminaire de géométrie algébrique du Bois-Marie 1963–64 (SGA 4): *Théorie des topos et cohomologie étale des schémas*, Lect. Notes in Math. **269, 270, 305**, Springer, 1972–73.

[SGA4$\frac{1}{2}$]   Séminaire de géométrie algébrique du Bois-Marie (SGA 4$\frac{1}{2}$): *Cohomologie étale*, Lect. Notes in Math. **569**, Springer, 1977.

[SGA5]   Séminaire de géométrie algébrique du Bois-Marie 1965–66 (SGA 5): *Cohomologie l-adique et fonctions L*, Lect. Notes in Math. **589**, Springer, 1977.

[SGA7]   Séminaire de géométrie algébrique du Bois-Marie 1965–66 (SGA 7): *Groupes de monodromie en géométrie algébrique*, Lect. Notes in Math. **288** et **340**, Springer, 1977.

[CorrGS]   P. Colmez, J.-P. Serre, eds. *Correspondance Grothendieck–Serre*, Doc. Math. **2**, SMF, 2001.

Articles

[1]   S. Abhyankar *On the field of definition of a nonsingular birational transform of an algebraic surface*, Ann. of Math. **66** (1957), 268–281.

[2]   Y. André *Pour une théorie inconditionnelle des motifs*, Publ. Math. IHÉS **83** (1996), 5–49.

[3]   Y. André, B. Kahn *Construction inconditionnelle de groupes de Galois motiviques*, C. R. Acad. Sci. Paris **334** (2002), 989–994.

[4]   Y. André, B. Kahn *Nilpotence, radicaux et structures monoïdales* (with an appendix by P. O'Sullivan), Rend. sem. mat. univ. Padova **108** (2002), 107–291.

[5]   Y. André, B. Kahn *Erratum: nilpotence, radicaux et structures monoïdales*, Rend. sem. mat. univ. Padova **113** (2005), 125–128.

[6]   A. Arabia, Z. Mebkhout *Sur le topos infinitésimal p-adique d'un schéma lisse*, I, Ann. Inst. Fourier **60** (2010), 1905–2094.

[7]   E. Artin *Über eine neue Art von L-Reihen*, Abh. Math. Sem. Univ. Hamburg **3** (1923), 89–108 = Collected Papers, 105–124, Addison-Wesley, 1965.

[8]   E. Artin *Beweis des allgemeinen Reziprozitätsgesetzes*, Abh. Math. Sem. Hamburg **5** (1927), 353–363.

[9]   E. Artin *Zur Theorie der L-Reihen mit allgemeinen Gruppencharakteren*, Abh. Math. Sem. Univ. Hamburg **8** (1930), 292–306 = Collected Papers, 165–179, Addison-Wesley, 1965.

[10]   E. Artin *Die gruppentheoretische Struktur der Diskriminante algebraischer Zahlkörper*, J. reine angew. Math. **164** (1931), 1–11.

[11]   E. Artin, G. Whaples *Axiomatic characterization of fields by the product formula for valuations*, Bull. AMS **51** (1945), 469–492.

[12]   M. Audin *La guerre des recensions. Autour d'une note d'André Weil en 1940*, Math. Semesterberichte **59** (2012), 243–260.

[13]   A. Beilinson, J. Bernstein, P. Deligne *Faisceaux pervers*, in Analyse et topologie sur les espaces singuliers (I), 6–10 juillet 1981, Astérisque **100**, SMF, 1982.

[14]   M. Bökstedt, A. Neeman *Homotopy limits in triangulated categories*, Compositio Math. **86** (1993), 209–234.

[15]   E. Borel *Sur une application d'un théorème de Hadamard*, Bull. Sc. Math. **18** (1894), 22–25.

[16] J. Bronowski *Curves whose grade is not positive in the theory of the base*, J. London Math. Soc. **13** (1938), 86–90.

[17] R. Brauer *On Artin L-series with general group characters*, Ann. Math **48** (1947), 502–514.

[18] C. Breuil, B. Conrad, F. Diamond, R. Taylor *On the modularity of elliptic curves over* **Q**: *wild 3-adic exercises*, J. AMS **14** (2001), 843–939.

[19] C. Chevalley *Généralisation de la théorie du corps de classes pour les extensions infinies*, J. Math. pures appl. **15** (1936), 359–371.

[20] J. B. Conrey *More than two fifths of the zeros of the Riemann zeta function are on the critical line*, J. reine angew. Math. **399** (1989), 1–16.

[21] G. Castelnuovo *Sulle serie algebriche di gruppi di punti appartenenti ad una curva algebrica*, Rom. Acc. L. Rend. (5) **15**, N° 1 (1906), 337–344. Reproduced in *Memorie scelte*, n° XXVIII, 509–517, Bologna, N. Zanichelli, 1937.

[22] Ph. Cassou-Noguès, T. Chinburg, A. Fröhlich, M. J. Taylor *L-functions and Galois modules* (from notes by D. Burns and N. P. Byott), London Math. Soc. Lect. Note Ser. **153**, *L*-functions and arithmetic (Durham, 1989), 75–139, Cambridge University Press, 1991.

[23] Ch. J. de la Vallée-Poussin *Recherches analytiques sur la théorie des nombres premiers*, Ann. Soc. Sci. Bruxelles **20** (1896), 183–256.

[24] P. Deligne *Les constantes des équations fonctionnelles*, Sém. Delange-Pisot-Poitou **11** (1969–70), exp. **19 bis**, 16–28.

[25] P. Deligne *Les constantes des équations fonctionnelles des fonctions L*, Modular functions in one variable II (Proc. Summer School, Univ. Antwerp, RUCA, July 17–August 3, 1972), Lect. Notes in Math. **349**, Springer, 1973, 501–597.

[26] P. Deligne *La conjecture de Weil, I*, Publ. Math. IHÉS **43** (1974), 273–307.

[27] P. Deligne *Théorie de Hodge, III*, Publ. Math. IHÉS **44** (1974), 5–77.

[28] P. Deligne *Les constantes locales de l'équation fonctionnelle de la fonction L d'Artin d'une représentation orthogonale*, Invent. Math. **35** (1976), 299–316.

[29] P. Deligne *Valeurs de fonctions L et périodes d'intégrales*, with an appendix by N. Koblitz and A. Ogus, Proc. Symp. Pure Math. **XXXIII 2**, Automorphic forms, representations and *L*-functions (Oregon State Univ., 1977), AMS, 1979, 313–346.

[30] P. Deligne *La conjecture de Weil, II*, Publ. Math. IHÉS **52** (1980), 137–252.

[31] P. Deligne, G. Henniart *Sur la variation, par torsion, des constantes locales d'équations fonctionnelles de fonctions L*, Invent. Math. **64** (1981), 89–118.

[32] A. Deninger, J. P. Murre *Motivic decomposition of abelian schemes and the Fourier transform*, J. reine angew. Math. **422** (1991), 201–219.

[33] B. de Smit, R. Perlis *Zeta functions do not determine class numbers*, Bull. AMS **31** (1994), 213–215.

[34] M. Deuring *Arithmetische Theorie der Korrespondenzen algebraischer Funktionenkörper, I, II*, J. reine angew. Math. **177** (1937), 161–191 and **183** (1940), 25–36.

[35] M. Deuring *Die Typen der Multiplikatorenringe elliptischer Funktionenkörper*, Abh. Math. Sem. Univ. Hamburg **13** (1940), 197–272.

[36] M. Deuring *Die Zetafunktion einer algebraischen Kurve von Geschlechte Eins*, II, Gött. Nachr., 1955, n° 2, 133–42; ibid., 1956, n° 4, 37–76. See also: *On the zeta-function of an elliptic function field with complex multiplication*, Proc.

Int. Symposium Alg. Numb. Theory (Tokyo-Nikko, 1955), Science Council of Japan, 1956, 47–50.

[37] G. Lejeune-Dirichlet *Beweis des Satzes, daß jede unbegrenzte arithmetische Progression, deren erstes Glied und Differenz ganze Zahlen ohne gemein-schaftlichen Faktor sind, unendlich viele Primzahlen enthält*, read at the Berlin Academy of Sciences on 27 June 1837, *in* G. Lejeune-Dirichlet's Werke, Band I, 313–342, Chelsea, New York, 1969. French translation: J. Math. Pures et Appl. (Journal de Liouville) (I) **4** (1839), 393–422.

[38] B. Dwork *On the rationality of the zeta function of an algebraic variety*, Amer. J. Math. **82** (1960), 631–648.

[39] G. Eisenstein *Beweis der allgemeinsten Reciprozitätsgesetzte zwischen reellen und complexen Zahlen*, Monatsber. d. preuss. Akad. d. Wiss. zu Berlin, 1850, 189–198.

[40] T. Ekedahl *On the adic formalism*, *in* The Grothendieck Festschrift (P. Cartier et al., eds.), Vol. II, Birkhäuser 1990, 197–218.

[41] K. Ford *Vinogradov's integral and bounds for the Riemann zeta function*, Proc. London Math. Soc. **85** (3) (2002), 565–633.

[42] A. Grothendieck *Sur une note de Mattuck–Tate*, J. reine angew. Math. **200** (1958), 208–215.

[43] A. Grothendieck *Formule de Lefschetz et rationalité des fonctions L*, Sém. Bourbaki **279**, 1964/65, *in* Dix exposés sur la cohomologie des schémas, (A. Grothendieck, N.H. Kuiper eds.), North Holland–Masson, 1968, 31–45.

[44] A. Grothendieck *Standard conjectures on algebraic cycles*, Algebraic Geometry (Internat. Colloq., Tata Inst. Fund. Res., Bombay, 1968) 193–199, Oxford University Press, 1969.

[45] J. Hadamard *Sur la distribution des zéros de la fonction $\zeta(s)$ et ses conséquences arithmétiques*, Bull. SMF **14** (1896), 199–220.

[46] J. Hall, D. Rydh *Perfect complexes on algebraic stacks*, Compositio Math. **153** (2017), 2318–2367.

[47] G.H. Hardy *Sur les zéros de la fonction $\zeta(s)$ de Riemann*, C. R. Acad. Sci. Paris **158** (1914), 1012–1014.

[48] H. Hasse *Theorie der relativ-zyklischen algebraischen Funktionenkörper, insbesondere bei endlichem Konstantenkörper*, J. reine angew. Math. **172** (1935), 37–54.

[49] H. Hasse *Zur Theorie der abstrakten elliptischen Funktionenkörper, I, II, III*, J. reine angew. Math. **175** (1936), 55–62, 69–88, 193–207.

[50] E. Hecke *Eine neue Art von Zetafunktionen und ihre Beziehungen zur Verteilung der Primzahlen (Zweite Mitteilung)*, Math. Z. **6** (1920), 11–51 = Mathematical Works 249–289, Vandenhoeck und Ruprecht, 1959.

[51] H. Heilbronn *On the class number in imaginary quadratic fields*, Quarterly J. Math. **5** (1934), 150–160.

[52] G. Henniart *Galois ε-factors modulo roots of unity*, Invent. Math. **78** (1984), 117–126.

[53] T. Honda *Isogeny classes of abelian varieties over finite fields*, J. Math. Soc. Japan **20** (1968), 83–95.

[54] J. Igusa *On the theory of algebraic correspondences and its application to the Riemann hypothesis in function-fields*, J. Math. Soc. Japan **1** (1949), 147–197.

[55] L. Illusie *Grothendieck et la cohomologie étale* (2008), *in* Alexandre Grothendieck: a mathematical portrait, 175–192, Int. Press, 2014.

[56] K. Iwasawa *On the rings of valuation vectors*, Ann. of Math. (2) **57** (1953), 331–356.

[57] U. Jannsen *Motives, numerical equivalence, and semisimplicity*, Invent. Math. **107** (1992), 447–452.

[58] J.-P. Jouanolou *Théorèmes de Bertini et Applications*, Progress in Math. **42**, Birkhäuser (1983).

[59] B. Kahn *On the semi-simplicity of Galois actions*, Rend. Sem. Math. Univ. Padova **112** (2004), 97–102.

[60] B. Kahn *Zeta functions and motives*, Pure appl. math. quarterly **5** (2009), 507–570 [2008].

[61] B. Kahn *Quelques calculs de sommes de Gauss*, Ann. Sci. Math. Québec **36** (2012), 487–500.

[62] B. Kahn *Motifs et adjoints*, Rendiconti Sem. Math. Univ. Padova **139** (2018), 77–128.

[63] B. Kahn *Zeta and L-functions of triangulated motives*, in preparation, cf. http://webusers.imj-prg.fr/~bruno.kahn/preprints/zetaL-Regensburg.pdf.

[64] B. Kahn *L'inégalité de Castelnuovo–Severi, (d')après Weil*, in preparation.

[65] B. Kahn, G. Maltsiniotis *Structures de dérivabilité*, Adv. in Math. **218** (2008), 1286–1318.

[66] B. Kahn, J.-P. Murre, C. Pedrini *On the transcendental part of the motive of a surface*, *in* Algebraic cycles and motives (workshop on the occasion of the 75th birthday of J.P. Murre), LMS Series **344** (2), Cambridge University Press, 2007, 143–202.

[67] E. Kani *On Castelnuovo's equivalence defect*, J. reine angew. Math. **332** (1984), 24–70.

[68] T. Katsura, T. Shioda *On Fermat varieties*, Tôhoku Math. J. **31** (1979), 97–115.

[69] N. Katz, W. Messing *Some consequences of Deligne's proof of the Riemann hypothesis for varieties over finite fields*, Invent. Math. **23** (1974), 73–77.

[70] K. Kedlaya *Fourier transforms and p-adic 'Weil II'*, Compos. Math. **142** (2006), 1426–1450.

[71] S. Kleiman *Algebraic cycles and the Weil conjectures*, *in* Dix exposés sur la cohomologie des schémas (A. Grothendieck, N.H. Kuiper eds.), North Holland–Masson, 1968, 359–386.

[72] S. Kleiman *Motives*, Algebraic geometry, Oslo, 1970 (F. Oort, ed.), Walters-Noordhoff, 1972, 53–82.

[73] S. Kleiman *The standard conjectures*, *in* Motives (U. Jannsen, S. Kleiman, J.-P. Serre, eds.), Proc. Symp. pure Math. **55** (1), 3–20.

[74] K. Künnemann *On the Chow motive of an abelian scheme*, *in* Motives (U. Jannsen, S. Kleiman, J.-P. Serre, eds.), Proc. Symp. pure Math. **55** (1), 189–205.

[75] E. Landau *Über einen Satz von Tschebyschef*, Math. Ann. **61** (1905), 527–550.

[76] S. Lang, A. Weil *Number of points of varieties in finite fields*, Amer. J. Math. **76** (1954), 819–827.

[77] R. P. Langlands *On the Functional Equation of the Artin L-functions*, unfinished notes (1970), http://publications.ias.edu/rpl/paper/61.

[78] G. Laumon *Transformation de Fourier, constantes d'équations fonctionnelles et conjecture de Weil*, Publ. Math. IHÉS **65** (1987), 131–210.

[79] S. Lichtenbaum *The Weil-étale topology on schemes over finite fields*, Compos. Math. **141** (2005), 689–702.

[80] Q. Liu, J. Tong, *Néron models of algebraic curves*, Trans. AMS **368** (2016), 7019–7043.

[81] Yu. Manin *Correspondences, motifs and monoidal transformations*, Math. USSR Sbornik **6** (1968), 439–470.

[82] J. Martinet *Character theory and Artin L-functions*, in Algebraic number fields: *L*-functions and Galois properties (Proc. Sympos., Univ. Durham, Durham, 1975) (A. Fröhlich, ed.), Acad. Press, 1977, 1–87.

[83] J. Milne *Abelian varieties*, in Arithmetic geometry (Cornell, Silverman, eds.), Springer, 1986 (rev. second printing, 1998), 103–150.

[84] J. Milne *Jacobian varieties*, in Arithmetic geometry (Cornell, Silverman, eds.), Springer, 1986 (rev. second printing, 1998), 167–212.

[85] T. Matsusaka *The criteria for algebraic equivalence and the torsion group*, Amer. J. Math. **79** (1957), 53–66.

[86] J. Mattuck, J. Tate *On the inequality of Castelnuovo–Severi*, Abh. Math. Sem. Univ. Hamburg **22** (1958), 295–299.

[87] A. Neeman *The Grothendieck duality theorem via Bousfield's techniques and Brown representability*, J. Amer. Math. Soc. **9** (1996), 205–236.

[88] L. B. Nisnevič *On the number of points of an algebraic manifold in a prime finite field* (in Russian), Dokl. Akad. Nauk SSSR (N.S.) **99** (1954), 17–20.

[89] B. Osserman *The Weil conjectures*, in Princeton Companion to Mathematics, Princeton University Press, 2008.

[90] S. Patrikis and R. Taylor *Automorphy and irreducibility of some l-adic representations*, Compos. Math. **151** (2015), no. 2, 207–229.

[91] C. Pépin *La formule des traces de Lefschetz–Verdier*, rencontre ARIVAF 4, univ. Bordeaux 1 (2010), exposé 4, www.math.polytechnique.fr/~cadoret/ARIVAF_Rencontre4_Cedric.pdf.

[92] R. Perlis *On the equation $\zeta_K(s) = \zeta_{K'}(s)$*, J. Number Theory **9** (1977), 342–360.

[93] M. Rapoport, T. Zink, *Über die lokale Zetafunktion von Shimuravarietten, Monodromiefiltration und verschwindende Zyklen in ungleicher Charakteristik*, Inv. Math. **68** (1982), 21–201.

[94] M. Raynaud *Caractéristique d'Euler–Poincaré d'un faisceau et cohomologie des variétés abéliennes*, Sém. Bourbaki 1964–66, exp. **286**, in Dix exposés sur la cohomologie des schémas, A. Grothendieck, N.H. Kuiper, eds, North Holland–Masson, 1968.

[95] B. Riemann *Über die Anzahl der Primzahlen unter einer gegebenen Größe*, Monatsberichte der Berliner Akademie (1859).

[96] J. Riou Théorie homotopique des *S*-schémas, mémoire de DEA, Univ. Paris 7 – Paris Diderot, 2002, www.math.u-psud.fr/~riou/dea/dea.pdf.

[97] P. Roquette *Arithmetischer Beweis der Riemannschen Vermutung in Kongruenzfunktionenkörpern beliebigen Geschlechts*, J. reine angew. Math. **191** (1953) 199–252.

[98]   P. Roquette *The Riemann hypothesis in characteristic p, its origin and development (I, II, III, IV,...)*, Mitt. Math. Ges. Hamburg **21, 23, 25, 32**.
        www.rzuser.uni-heidelberg.de/~ci3/manu.html.

[99]   M. Saito *Monodromy filtration and positivity*, preprint, 2000,
        http:// arxiv.org/abs/math/0006162

[100]   T. Saito *Weight spectral sequences and independence of l*, J. Inst. Math. Jussieu
        2 (2003), 583–634.

[101]   F.K. Schmidt *Analytische Zahlentheorie in Körpern der Charakteristik p*, Math.
        Zeit. **33** (1931), 1–22.

[102]   C. Schnell *An overview of Morihiko Saito's theory of mixed Hodge modules*,
        http://arxiv.org/abs/1405.3096, to appear in Tsinghua Sanya Int. Math. Forum.

[103]   L. Schoenfeld *Sharper bounds for the Chebyshev functions $\theta(x)$ and $\psi(x)$, II*,
        Mathematics of Computation **30** (134) (1976), 337–360.

[104]   A. Scholl *Classical motives, in* Motives (U. Jannsen, S. Kleiman, J.-P. Serre,
        eds.), Proc. Symp. pure Math. **55** (1), 163–187.

[105]   P. Scholze *Perfectoid spaces*, Publ. math. de l'IHÉS **116** (2012), 245–313.

[106]   B. Segre *Intorno ad teorema di Hodge sulla teoria della base per le curve di
        una superficie algebrica*, Ann. Mat. **16** (1937), 157–163.

[107]   J.-P. Serre *Groupes d'homotopie et classes de groupes abéliens*, Ann. of Math.
        58 (1953), 258–294 = Oeuvres – Collected Papers, vol. I, 171–207.

[108]   J.-P. Serre *Faisceaux algébriques cohérents*, Ann. of Math. **61** (1955), 197–278
        = Oeuvres – Collected Papers, vol. I, 310–391.

[109]   J.-P. Serre *Analogues kählériens de certaines conjectures de Weil*, Ann. of
        Math. **71** (1960), 392–394 = Oeuvres – Collected Papers, vol. II, 1–3.

[110]   J.-P. Serre *Rationalité des fonctions ζ des variétés algébriques (d'après B.
        Dwork)*, Sém. Bourbaki, 1959/60, exp. n° 198 = Seminar papers 1950–1999,
        Doc. Math. **1**, SMF, 2001, 169–178.

[111]   J.-P. Serre *Endomorphismes complètement continus des espaces de Banach
        p-adiques*, Publ. Math. IHÉS **12** (1962), 69–85 = Oeuvres – Collected Papers,
        vol. II, 170–186.

[112]   J.-P. Serre *Zeta and L-functions, in* Arithmetical algebraic geometry, Harper
        and Row (1965), 82–92 = Oeuvres – Collected Papers, Vol. II, 249–259.

[113]   J.-P. Serre *Facteurs locaux des fonctions zêta des variétés algébriques: définitions et conjectures*, Sém. Delange-Pisot-Poitou, 1969–70 = Oeuvres –
        Collected Papers, vol. II, 581–592.

[114]   J.-P. Serre, J. Tate *Good reduction of abelian varieties*, Ann. of Math. **88** (1968),
        492–517, = Oeuvres – Collected Papers, vol. II, 492–497.

[115]   J.-P. Serre *Motifs*, Journées arithmétiques, Luminy, *in* Astérisque **198–199–200**
        (1991), 333–349 = Oeuvres – Collected Papers, vol. IV, 223–239.

[116]   J.-P. Serre *La vie et l'œuvre d'André Weil*, L'Ens. math. **45** (1999), 5–16 =
        Oeuvres – Collected Papers, vol. IV, 653–664.

[117]   F. Severi *Sulla deficienza della serie caratteristica di un sistema lineare di curve
        appartenenti ad una superficie algebrica*, Atti Accad. Naz. Lincei, Rend. **12**,
        1903.

[118]   C. L. Siegel *Über die Classenzahl quadratischer Zahlkörper*, Acta arith. **1**
        (1935), 83–86.

[119] J. Shurman *Hecke characters, classically and idèlically*, http://people.reed.edu/~jerry/361/lectures/heckechar.pdf.

[120] N. Spaltenstein *Resolutions of unbounded complexes*, Compositio Math. **65** (1988), 121–154.

[121] J. Stallings *Centerless groups – an algebraic formulation of Gottlieb's theorem*, Topology **4** (1965), 129–134.

[122] H. M. Stark *Dirichlet's class-number formula revisited*, A tribute to Emil Grosswald: number theory and related analysis, 571–577, Contemp. Math. **143**, AMS, Providence, RI, 1993.

[123] R. G. Swan *K-theory and algebraic correspondences*, *in* Proceedings of the Conference on Orders, Group Rings and Related Topics (Ohio State Univ., Columbus, Ohio, 1972), Lect. Notes in Math., **353**, Springer, 1973, 161–179.

[124] L. Szpiro *Degrés, intersections, hauteurs*, *in* Séminaire sur les pinceaux arithmétiques: la conjecture de Mordell, Astérisque **127** (1985), 11–28.

[125] K. Takagi, *Über das Reciprocitätsgesetz in einem beliebigen algebraischen Zahlkörper*, J. College of Science, Imp. Univ. of Tokyo **44** (1922), 1–50.

[126] J. Tate *Fourier analysis in number fields, and Hecke's zeta-functions*, doctoral thesis, Princeton, 1950. Reproduced *in* Algebraic Number Theory (Proc. Instructional Conf., Brighton, 1965), J.W.S. Cassels, A. Fröhlich, eds., Acad. Press, Thompson, 1967, 305–347.

[127] J. Tate *Endomorphisms of abelian varieties over finite fields*, Invent. math. **2** (1966), 134–144.

[128] J. Tate *The general reciprocity law*, *in* Mathematical developments arising from Hilbert problems (F. Browder, ed.), Proc. Symp. pure Math. **28** (2), AMS, 1976, 311–322.

[129] J. Tate *Local constants* (prepared in collaboration with C. J. Bushnell and M. J. Taylor), *in* Algebraic number fields: *L*-functions and Galois properties (Proc. Sympos., Univ. Durham, Durham, 1975) (A. Fröhlich, ed.), Acad. Press, 1977, 89–131.

[130] J. Tate *Conjectures on algebraic cycles in ℓ-adic cohomology*, *in* Motives (U. Jannsen, S. Kleiman, J.-P. Serre, eds.), Proc. Symp. pure Math. **55** (1), 71–83.

[131] R. Taylor, A. Wiles *Ring-theoretic properties of certain Hecke algebras*, Annals of Math. **141** (1995), 553–572.

[132] T. Terasoma *Monodromy weight filtration is independent of l*, preprint, 1998, http://arxiv.org/abs/math/9802051.

[133] R. Thomason, T. Trobaugh *Higher algebraic K-theory of schemes and of derived categories*, The Grothendieck Festschrift, Vol. III, Progr. Math., **88**, Birkhäuser, 1990, 247–435.

[134] V. Voevodsky A^1-*homotopy theory*, *in* Proc. Int. Congress of Mathematicians (Berlin, 1998), Documenta Mathematica, Extra Vol. I (1998), 579–604.

[135] V. Voevodsky *Triangulated categories of motives over a field*, *in* Cycles, transfers, and motivic homology theories, Ann. of Math. Stud. **143**, Princeton University Press, 2000, 188–238.

[136] H. von Koch *Sur la distribution des nombres premiers*, Acta Math. **24** (1901), 159–182.

[137] F. Wecken *Fixpunktklassen*, I: Math. Ann. **117** (1941), 659–671, II, III: *ibid.*, **118** (1942–43), 216–234 and 544–577.

[138] A. Weil *Sur les fonctions algébriques à corps de constantes fini*, C. R. Acad. Sci. Paris **210** (1940), 592–594 = Œuvres Scientifiques: Collected Papers, vol. I, [1940b].

[139] A. Weil *On the Riemann hypothesis on function fields*, Proc. Nat. Ac. Sci. **27** (1941), 345–347 = Œuvres Scientifiques: Collected Papers, vol. I, [1941].

[140] A. Weil *Numbers of solutions of equations in finite fields*, Bull. AMS **55** (1949), 497–508 = Œuvres Scientifiques: Collected Papers, vol. I, [1949b].

[141] A. Weil *Sur la théorie du corps de classes*, J. Math. Soc. Japan **3** (1951), 1–35 = Œuvres Scientifiques: Collected Papers, vol. I, [1951b].

[142] A. Weil *Sur les "formules explicites" de la théorie des nombres premiers*, Comm. Lund, (1952) (volume dédié à Marcel Riesz) = Œuvres Scientifiques: Collected Papers, vol. II, [1952b].

[143] A. Weil *Jacobi sums as "Größencharaktere"*, Trans. AMS **73** (1952), 487–495 = Œuvres Scientifiques: Collected Papers, vol. II, [1952d].

[144] A. Weil *Sur les critères d'équivalence en géométrie algébrique*, Math. Ann. **128** (1954), 95–127 Œuvres Scientifiques: Collected Papers, vol. II, [1954d].

[145] A. Weil *On a certain type of characters of the idèle class group*, Proc. Int. Symp. on algebraic number theory, Tokyo-Nikko, 1955, 1–7 = Œuvres Scientifiques: Collected Papers, vol. II, [1955c].

[146] A. Weil *Über die Bestimmung Dirichletscher Reihen durch Funktionalgleichungen*, Math. Ann. **168** (1967), 149–156 = Œuvres Scientifiques: Collected Papers, [1967a].

[147] A. Weil *Sommes de Jacobi et caractères de Hecke*, Gött. Nachr. (1974) = Œuvres Scientifiques: Collected Papers, vol. III, [1974d].

[148] A. Wiles *Modular elliptic curves and Fermat's last theorem*, Annals of Math. **141** (1995), 443–551.

Index

Printed in the United States
by Baker & Taylor Publisher Services